VENICE DESIGN
2019

VENICE DESIGN 2019

COLOPHON

© 2019. Texts by the authors.
© Unless otherwise mentioned, photos courtesy of the designers and the European Cultural Centre.

All rights reserved. No part of this publication may be reproduced, stored in retrieval system, or transmitted in any form or by any means, electronic, mechanical, photocopying, recording or otherwise, without permission of the editor. Although the editors and publisher have made every effort to ensure that the image rights have been arranged and accurately credited, the author and publisher do not assume and hereby disclaim any liability.

www.venice-design.com
www.europeanculturalcentre.eu

ISBN 9789 0829 434 12

PUBLISHED BY
European Cultural Centre
GRAPHIC DESIGN AND EDITING
ROSE Design & Communication
PRINT
Elcograf S.p.A., Italy
ACKNOWLEDGEMENTS
Debora Bae, Tiziano De Lazzari, Giorgia De Santi, Jhan Thomas Montgomery, Eugenio Pettirossi and Micaeala Skerl

CONTENTS

INTRODUCTION 8

DESIGNERS

&SOCIETY	14	
Farah Abdelhamid	16	
acoocooro	18	
ALP	20	
Atelier Areti	22	
Atelier Nick Boers	24	
Masayo Ave	26	
Alexander Bannink Industrial Design & Exit International	28	
Lacy Barry	32	
Nina van Bart	34	
Donald Baugh	36	
BC Biermann	38	
Loreta Bilinskaite-Monie	40	
Felicia Björklund	42	
Daniela Buonvino	44	
Kevin Callaghan	46	
Barbara Calvo Design	48	
Piero Castiglioni	50	
Sophia Chraïbi Giorgi	52	
cloudandco	54	
Luce Couillet	56	
Creative Chef	58	
Dá Design Studio	60	
Saverio D'Elia	62	
DUODU	64	
Edinburgh College of Art	66	
FAINA Collection	68	
Hicham Ghandour	70	
GHYCZY	72	
Donna Glubo-Schwartz	74	
Julie Helles Eriksen	76	
Barrie Ho	78	
Joca van der Horst	84	
Hsc Designs	86	
Katalin Huszár	88	
Muraad Ibrahim	90	
Megumi Ito	92	
Studio Silvia Knüppel	94	
Eva Levin	96	
Lumneo	98	
Lyk Carpet x Unique Factory	100	
Tahir Mahmood	102	
M.I.C.A.T	104	
Teresa Moorhouse	106	
Brian Naeyaert	108	
Nivasarkar Consultants	110	
Jinhwa Oh	112	
Pemara Design	114	
PY MANUFACTURE	116	
Qstudio	118	
REVOLOGY	120	
Lou van 't Riet	122	
Selek	124	
Gisela Simas	126	
Mikaela Steby Stenfalk	128	
J.M. Szymanski & A. Kohl	130	
S. Tick, M. Wallis & R. Roepnack	134	
Chantal Tramasure	136	
Ritu Varuni	138	
Wearable Media	140	
worktecht	142	
Jungmo Yang	144	
Your Artist	146	
Monika Zabel	148	
Hongtao Zhou, Fang Qi, Abaiyi-Akebai & Lianglu Hou	150	

MADE IN VENICE 156

CURATORS 182

IMAGE CAPTIONS 186

SPONSORS 190

INTRODUCTION

INTRODUCTION

THE EUROPEAN CULTURAL CENTRE

The European Cultural Centre, a place for reflection, research, and creation for interdisciplinary encounters, provides the conditions to invite artistic and creative practices from all fields since 2011 – visual art, dance, performance, music, literature, architecture…,seeing them as a process of learning and experiencing.

Indeed, the ECC reflects upon the dynamics of European culture and influences, upon how Europe is seen within and outside its borders. Its aim is to go beyond our geographical borders. Borders – in the widest sense of the word – have to be crossed in order to develop ourselves as human beings. "To cherish our differences and strengthen cultural commons", this goal can only become reality if we open ourselves to the world around us and share our thoughts, without prejudice.

With VENICE DESIGN as youngest addition to ECC's activities, the ECC has brought and opened up a wider Design field in the city of Venice during the Biennial period.

For hundreds of years, Venice has been a place of cultural exchange and an important exporter of European culture. Venice is a city with an extraordinary concentration of facilities and organizations dedicated to culture, which makes it the ideal venue for the realisation of the objectives of the ECC. The historic centre is only populated by approx. 60.000 inhabitants, but it sustains: 45 museums and 7 theatres, 14 foundations with the objective to promote culture; 2 leading universities, an art academy, a conservatory, and many public libraries; 32 consulates and regional offices of a.o. UNESCO, WHO, and the Council of Europe; and it is also the city of La Biennale di Venezia.

VENICE DESIGN 2019

Technological advances, lack of potable water, excessive plastic consumption, alternative ways of producing or communicating, improvement of well-being… From humans' physiological needs to digital practices, designers' reflections and researches nowadays have no limits. More and more comprehensive and overflowing, the design field cannot overlook the changes our society is facing today.

When exploring all aspects of our ways of living, Designers are increasingly embracing the essence of designing - a combination of meaning, functionality and senses. Acting in both large and small scales, creators are proposing new and unconventional ways of reconnecting with the sensory faculty and of experiencing the different perceptions and realities.

In line with this premise, the fourth edition of VENICE DESIGN gathers 60 Designers from 30 different countries. Presented in the iconic venue of Palazzo Michiel, the show aims to share experiences and investigations on the role of Design in today's society.

The exhibition invites the audience to discover cutting-edge design concepts, as well as creative solutions for daily life uses and consumption patterns. Practices involving new technologies will be presented next to tributes to tradition and craftsmanship by highlighting the conception and production processes of the projects. In an engaging and creative environment, the visitor plays a central role by becoming the protagonist and by interacting with the pieces. Physically engaged, the visitor is stimulated to think of important contemporary themes such as our well-being, our production systems, and our means of communication.

Initiating with a reflection upon our most basic needs (eating, drinking…), Designers are paying genuine attention to natural resources' accessibility and availability, as well as rethinking our standard behaviours.

The project of **Yeongkyu YOO X Coway**, for instance, questions the excessive bottled water consumption and proposes an alternative for plastic usage. On a different note, **Creative Chef Studio**, a project led by Jasper Udink ten Cate, composes happenings with food and design objects, creating a different context for the vital act of eating.

Designers focus on enhancing our relationship with what surrounds us, not only in a palpable sense but also on an emotional level. While aesthetically appealing, projects are made with consciously chosen materials and produced in an ethical and respectful way towards the environment. In fact, sustainable and eco-conscious approaches are ubiquitous in today's practices.

Inspired by authentic local traditions, creators such as **Sophia Chraïbi Giorgi** conceive furniture by combining an eco-friendly method with the manufacturing process. Others as **FAINA Collection**, **GHYCZY** and **Shrikant Nivasarkar** are mindful of ecological issues, while embracing their cultures and demonstrating a unique savoir-faire. A special care for materials is evident in the practices of Designers such as **Suzanne Tick**, who presents an innovative process for hand-weaving by creating compositions with mediums like Mylar® balloons and neon lights. Combining 3D printed recycled plastic and MUJA traditional hand-weaving techniques, the consortium formed by **Aziza Chaouni Projects** and **IFASSEN** demonstrates a commitment to sustainability, environmental protection and social impact.

Today, innovation is inevitably related to new technologies. Part of our daily life, they are invading our public and intimate spheres, modifying our behaviors and redefining our needs. In the field of Design, their impact has been crucial in many aspects, especially when it comes to broadening the possibilities of creating new experiences for users.

Focusing on the new media subculture of video games and its community, **Barrie Ho Architects** fully immerse the audience into their latest project "E-Sport Stadium", the first-ever build arena exclusively dedicated to gamers. With a different approach, the visionary studio **Wearable Media** offers new physical and social experiences by integrating into their fashion creations sound interaction technology, and even artificial intelligence.

The effects of new technologies are not limited to fleeting moments of experience, they actually are changing our perceptions of reality. With the help of devices and gadgets, one can travel and discover other parts of the globe. Projects such as the site-specific installation of **BC Bierman** use augmented reality to virtually reproduce real-world environments.

By conceiving and designing experiences, creators envision interactions. Those are moments in which the visitors reconnect with their sensations and perceptions by performing the role of activating the pieces.

Responding to the movements produced by the visitors, the interactive installation of **Daniela Buonvino** comes alive. Alternatively, **Farah Abdelhamid**, by putting the body in the core of her performative pieces, allows a deeper exploration of senses.

Ultimately, physical stimuli can lead to spiritual reflection as proposed in the project by **Acoocooro**. Famous for challenging the expected function of "common" objects, the multidisciplinary studio proposes an emotional experience transcending one's preconceived beliefs and rituals.

In addition, in a special collaboration with the company **Barrisol**, VENICE DESIGN 2019 is pleased to announce the presentation of iconic Italian Designer **Piero Castiglioni**'s Lamp "1954".

Finally, for the third year, the **VENICE DESIGN city map** will guide visitors to wander around the city and to discover 42 independent craftsmen living and working in the area. This ongoing collaboration between VENICE DESIGN and Venetian Designers aims at promoting the local Design scene's actors and traditions.

DESIGNERS

&SOCIETY

We are &SOCIETY, a design company performing various design activities with the theme of a mirror.

A Mirror is an objet where one faces the most honest "oneself".

We hope many people face themselves in the most beautiful places.

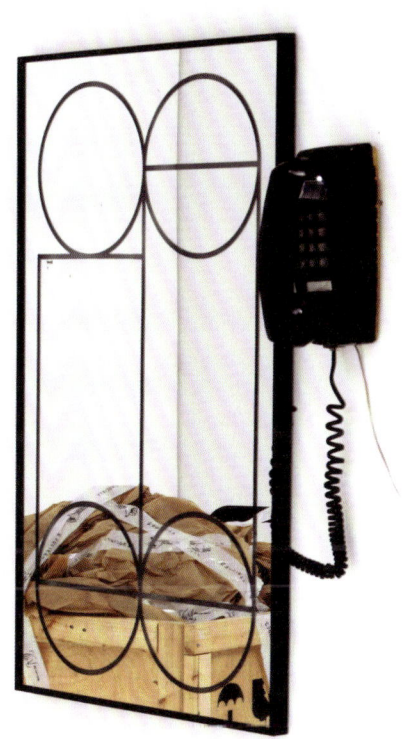

FARAH ABDELHAMID

IN/OUTROSPECTION is exclusively-made for VENICE DESIGN this year, featuring two playful installations, Union and In-Pression that investigate common jewelry theory, asking how and why does the body interact with a jewelry piece; what forms and materials heighten the awareness of our bodies; can we sense/see the moment of interaction; how the body exist as as subject and medium, and can an object redefine/direct the body or vice versa.

Through studying jewelry and the interconnectedness between functionality, material and scale, Farah began to reduce all pre-determined associations around jewelry to focus purely on the visual and sensory qualities of wearing, immersing the audience into a jewelry experience.

Union and *In-Pression* present two frameworks of jewelry design, using silicone rubber as its main material to emphasize its mass, weight, flexibility and fragility, color and scale, similar to that of the body and skin. One offers linear, heavy, dense hand-held and recognizable chain links composing a 5 meter necklace that can be "worn" in any variation depending on the person/s and their movement. The other presents 50+ soft pillow-like spheres that are like skin-thin casings mounted and protruding from a doorway, mimicking the exact moments that the skin morphs to accommodate the piece of jewelry as it is being worn.

Together, the two pieces offer the context of a playground, inviting the audience to touch, wear, carry, push, squeeze the pieces, together or individually, alternating their role between a participant and/or spectator in the framework of jewelry and wearability. The sensory results becomes not only internal but also external and visible as the wearer starts to move and accordingly, the silicone rubber, like skin and in similar neutral colors, reacts as well - hence "IN/OUTROSPECTION" which is the consequence of this engagement for the piece and the body alike.

Almost serving as social works to understand the body, its triggers, movements, and reactions, these pieces dissolve the theoretical obsession with aesthetic jewelry, and morph it into an awareness of the body, engaging the intellectual and physical participation of the audience.

Farah is an Egyptian jewelry artist. She received her formal training in Jewelry and Metalsmithing from the Rhode Island School of Design. Since then, Farah has been showcasing her commercial and conceptual jewelry at fairs and exhibitions in Amsterdam, Dubai, Egypt, and London with this year at Milan Design Week and Venice Design. Moreover, Farah has taught over 5 years of technical jewelry making and design courses and workshops, and is now building a community through her personal and shared studio space in Old Cairo, "StudiowithFarah".

ACOOCOORO

CEREMONIA

Design, objects, spaces, and images occupy our minds, fill our lives and shape every single aspect of our reality, the way we live and perceive it. Our daily interactions with ourselves, with others, and with the whole of human society are dictated by the actions and rituals that the objects and spaces that frame them trigger and conduct. With this in mind, human relationships with these objects and spaces cease to be circumstantial, and become as significant and consequential as human relationships themselves, hence the importance of revaluing rituals around things and reinstating the ceremonious in everyday life.

Objects must transcend the constraining and empty dimensions of mere function –where usefulness, advantage, and competence confer and deny intrinsic value–, and become something else: emotional symbols, playthings, containers and modifiers of history and memory; and experiences, that enrich both our inner and our public lives. *Ceremonia* is a continuation of this idea, an exploration on utilitarian objects, their relevance; the performances and rituals that surround them, informed by the history, music and reality contained within the old dance known in Mexico as *Concheros*. It is a collection, an experience where each object plays a part, pulling the viewer into the literal and symbolic performances that surround a specific configuration of "things" that would be construed differently if presented separately or in a different setting: performance is key, and meaning is found through ceremony.

Concheros is the name given to a ceremony performed in Mexico since the colonial period, along with the music and dance that accompany it. Syncretic in nature, it features both prehispanic and Christian elements, and its exact origins are difficult to connect to a specific prehispanic religious ceremony. Nowadays, however, it is performed during a number of Christian festivities, and many variants currently exist.

It is a circular and cyclical dance, performative, yet not intended for an audience. Calling upon a deity and their ancestors, concheros organize themselves into groups and hierarchies, and dance and play, in a ritual that brings past and present together. The sound of armadillo-shell mandolins, ayoyotes (shells, tied to the ankles), and drums fills the air, as fragrant herbs purify the space and those who take part.

Atmósfera, *Caja de música* and *Copaleras* bring the sounds, images, and mysticism of this rite into the realm of everyday presence, bodies, and entities: a lamp is no longer a lamp, music is more than a simple pastime, and a censer shows evidence of something more magical.

COLLABORATORS:

Martín Diego Salido-Orcillo, Antonio Mendoza, Andrés Aguilar, Marcos Pääp Gutiérrez.

ALP
ANNICK L PETERSEN

The alp design philosophy is to create simple, elegant products designed to last. I like to experiment with shapes creating a range of possibilities and a variety of ways to set the objects. Giving the owner of the object the choice of creating different settings depending on his/her need is an important part of the design.

The woven leather range from alp uses beautiful materials and expert craftsmanship. Combining a wonderfully tactile leather seat with angled steel base, the alp stool 104 has a bold contemporary shape that creates a strong feature. The stools come in two alternating shapes so they can be used individually or configured into a bench. I wanted to create a strong architectural shape for the base that would contrast with the softer leather top, the slim black metal also has mid-century influences. I love the fact you can create different shapes depending on the configuration of the stools.

The alp light 310 is made from black leather-like cord that is carefully stretched and knotted over a metal frame to create a dramatic geometric web. Playing with patterns and geometric shapes is a constant influence, inspiration comes from observing my surroundings, being a dress someone is wearing in the street, plants and nature or a brick pattern design on a building. They all play an important part and help me come up with new ideas and concepts.

ATELIER ARETI

Atelier Areti is an interdisciplinary design studio established by sisters Gwendolyn and Guillane Kerschbaumer. Gwendolyn and Guillane's background is in visual arts with a focus on drawing and sculpture, architecture and design.

Their work reflects this interest in both the object and space, exploring and existing between the sculptural quality of the object and its spacial dynamic.

We strive to develop something new and beautiful through our work. We are driven by curiosity and the desire to explore works that touch us on a deeper, direct emotional level as well engage us in a more conscious conceptual way. These artistic ambitions need to be realized and developed in the actual material product.

The material realization gives the design it's place in our daily life. A well done object will be valued for a long time; we aim to create pieces that remain relevant through their beauty, function and quality.

The square floor / wall light explores the regularity of the grid with the seeming randomness of the globes as well as the linearity of the tubes juxtaposed to the volume of the spheres.

Birch standing black + globe is a series based on two elements: a tube with irregular cuts that allow for light to come through the solid metal tube, and a white globe. This simple element is used in the floor light, but also in pendants, table lights and wall lights. It is used as a single element or in a row. Its simple shape appears solemn through the use of two simple solids, but at the same time has an element of lightness through the filigran and irregularly placed cuts.

The perspective series has as central element 4 spheres of different size, which aligned create the effect of the spheres receding in space.

The double sphere pendant is composed of two 'perspective' elements mirrored along a central vertical axis. It is a large piece with the two large globes at the bottom measuring 30 cm each in diameter.

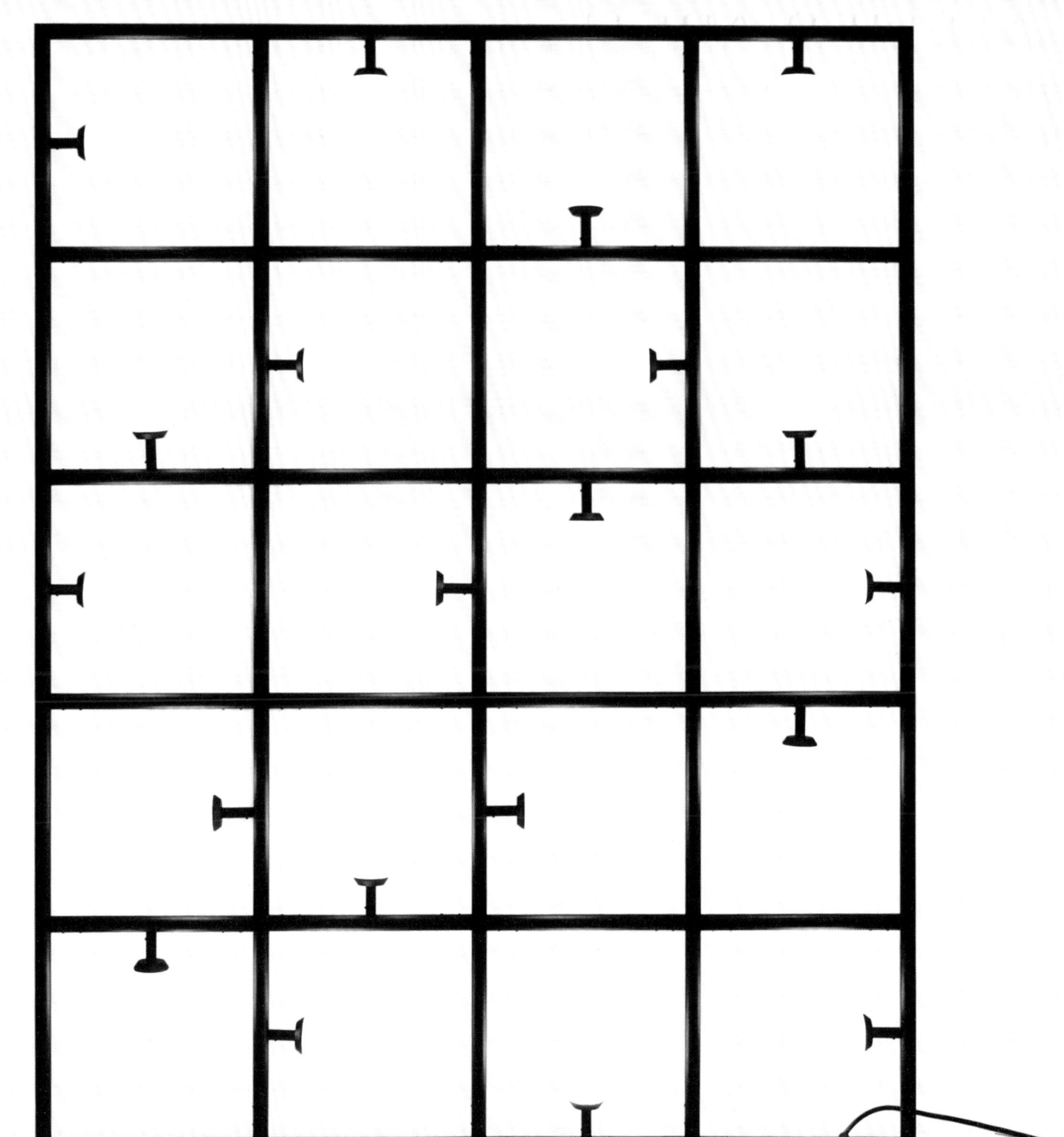

ATELIER NICK BOERS

In a humoristic and absurdist way the SPONGE series is trying to show my vision on the environmental problems we are dealing with today, focusing first on the troubles of giant garbage patches floating around in our seas globally.

SPONGE is the first visualization of the transformations our living organisms could go through if they were to adapt more to their surroundings, their surroundings becoming more and more filled with human products the organisms are starting to take shape and color that look very familiar to us.

The objects in this series are made from, what I like to call, a fictive narrative material, a material that was made up to tell a story. I found this to be an interesting way to approach a global issue without pointing any fingers or being very critical towards anyone.
For me this series could be the start of a whole database of fictive materials that each in their own way tell a story.

SPONGE tells the story of a sea sponge that is surrounded by trash for years and years and finally decides that if you cant beat them, just join them. So in a transformation that took decades, the sea sponge is gradually taking over the globally know yellow and green colors of the scotch bright cleaning sponge that floats around in all seas worldwide.

The colors of the cleaning sponge are so well know to everyone all around the world that it seemed to me to be a good way to get the story clear to a wide range of audience, from here on the story of SPONGE will be clear but it will also pave the way for upcoming objects made with new and different fictive narrative materials.

MASAYO AVE

What is Design? If somebody asks, I would reply;

> *Design is a continuous process of discoveries, which deals with colours, patterns, forms, structures, and its relationship perceived in everyday life. It is a profound multi-sensory experience that needs to cultivate from a very young age.*

Bringing the best of my three-decades-long expertise in design, I have been developing new design educational programs for everyone; not only for design students but also for children and youth as well as for the experts of many different fields. By careful planning the particular discovery process, I wish to grant everyone intuitive-, experiential and holistic understanding of the everyday environment.

DESIGN GYMNASTICS is a series of basic design exercise I developed to let one be mindful of everyday phenomena in their living environment. It is a sort of "treasure-hunting" play to discover "beautiful geometric shapes" designed by nature.

In Venice Design 2019, I present the best of the treasures hunted in the exercise of DESIGN GYMNASTICS A.B.C., which I practiced every day for ten years. It is the exercise to collect the complex of geometric shapes in the form of alphabet letters or numbers hidden in the details of humble nature in the city.

You can find the treasures not only in well-maintained green the city park but also in a forgotten corner of a parking lot. You can discover a fantastic letter inside a flower, once you look into it carefully with a magnifying lens. The dried leaves scattered on the ground sometimes form itself amazingly complex shape of a letter as a miracle of an instant. There exists a subtle treasure that grows silently beside your step. There exists a trampled treasure. There exists a crumpled treasure. There exists a frozen treasure. There exists a tiny number covered with fine hair that you can see only through a microscopic lens, and there exists a huge alphabet floats in the sky when you look up. The beautiful treasures do not always present their faces in front. Most of them lost themselves in a lot of messy things. They exist and just are waiting to be discovered - as if the answers in front of the eye but never noticed.

Learning the fundamentals of design is to get a comprehensive understanding of one's own living environment through a perceptive eye with the skills to evaluate its quality.

ALEXANDER BANNINK INDUSTRIAL DESIGN & EXIT INTERNATIONAL

THE CONCEPT
What if we had more than mere dignity to look forward to on our last day on this planet? What if we dared to imagine that our last day might also be one of our most exciting?

BACKGROUND
In 1996, Philip Nitschke became the first doctor in the world to give a legal, lethal voluntary injection under Australia's *Rights of the Terminally Ill Act*. Rather than administer the lethal dose himself, he created the Deliverance Machine computer to enable his patients to push the final button. More than 20 years later comes the 'Sarco' (or 'Pegasos'). Sarco is the first 3D-printed euthanasia capsule to allow a *elective*, peaceful and even euphoric death, while creating a sense of travel to a new destination. A lawful, non-medical alternative.

THE DESIGN
As an industrial designer working in automotive transportation, Alex Bannink's approach to the styling of Sarco has been inspired by the car. Sarco uses asymmetric design to guide entry while tapping the subconscious with an unexpected promise of adventure.

With its upward sweeping lines, Sarco is brought optically: a trick of the eye pointing you towards your end. In its stillness and serenity, Sarco presents the possibility of a journey that is - quite literally - dazzling. Sarco's interior reflects a bespoke vehicle, where the 'driver' is luxuriously seated in clear reach of the controls. This provides the ultimate front row seat to our collective destiny.

THE MECHANICS
The capsule of Sarco provides for a rapid decrease in oxygen level while maintaining a low CO_2 level. On activation, liquid nitrogen causes the oxygen level to drop silently to less than five percent. On inspiration, a sense of serene euphoria is followed quickly by loss of consciousness and peaceful death.

THE COLLABORATION
When Philip Nitschke left his home in Australia for a new working life in the Netherlands, his network brought him into contact with Dutch designer Alex Bannink. Sarco represents a nexus between the design aesthetics of modern transportation with the brief of delivering the ultimate endgame: a peaceful, elegant and stylish death at a time and place of one's choosing.

Far from shrouding death behind a grim, dark curtain, Sarco invites one's final day to be an overt display of beauty.

THE BUILD
Sarco is 3D-printed in sections measuring 1000x500x500mm, beginning with the frame: much like building a car. The robotized printer works 24/7 systematically manufacturing the main structure, body panels, and details' components. The nostalgia of hands-on production comes post-printing.

The 1:1 scale Sarco was printed over a period of six months using 160 kgs of biodegradable plastic.

THE FUTURE
In 2020, it is expected that Sarco will become a legitimate option for people from around the world who travel to Switzerland for a lawful, assisted death. By 2030, when large-scale 3D printing is easily accessible and older generations are living longer but sicker lives, Sarco will have practical application globally.

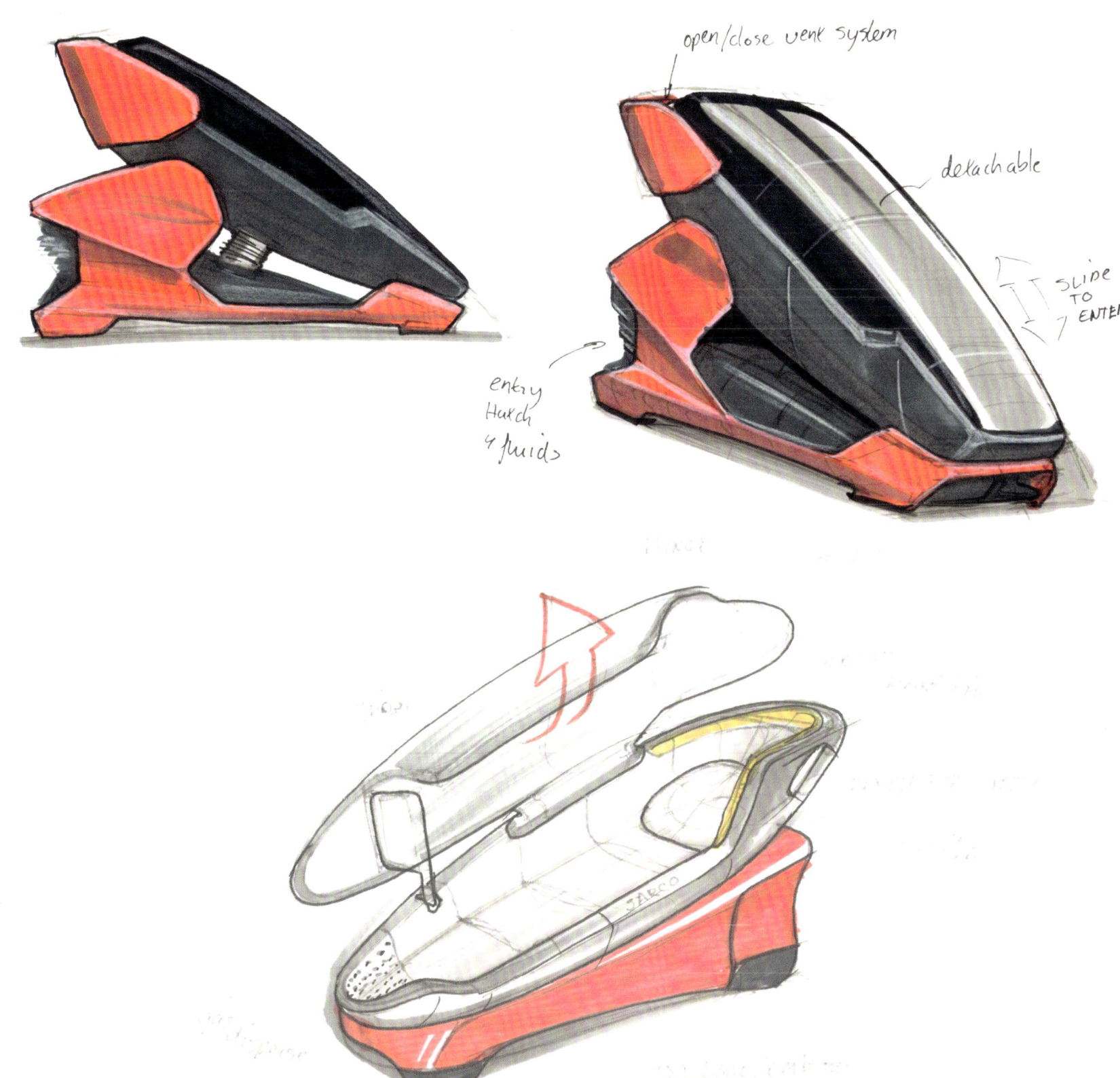

LACY BARRY

Most people at some point in their lives have felt displacement. In working with discarded materials —cardboard and paper— I explore displacement creatively by giving new purpose to the often under-appreciated materials that surround us everyday. I collect cardboard boxes—it seems weird, but once you find a beauty and a new use in a mundane material, it becomes valuable to you. To me in my practice, it is a durable yet incredibly versatile material to work with. Folding and cutting into it similar to paper, I can make just about anything with it. In this work I combine cardboard and paper into a two-piece sculpture aglow with colour. An intricate outer shell of paper serves as the 'aesthetic skin' that is fixed over a frame of cardboard 'bones'. It is painted, folded, curled and duplicated, giving a new placement to an otherwise displaced combination of materials. I started creating these works as a meditative process. Their presence gives me as much fulfillment and sanctity as I give them, and often I see personal feelings reflected in them that I cannot express in words. It's a contemplative journey I take pleasure in embarking on, evolving throughout the construction of a piece. This piece is the first duo of a series of pillar structures.

PILLAR OF SILVER CLOUD (BY DAY)

Every cloud has a silver lining, they say. Like the sun kissing the cloud's edge, making it glow, *Pillar of Silver Cloud* represents looming darkness with a glimmer of hope. A 'pillar of cloud' was also used scripturally to guide the lost to the promised land. Lush features of flowers, leaves, and intricate shapes invoke the abundance imagined at the end of that journey.

PILLAR OF GOLD FIRE (BY NIGHT)

Gold is refined by fire, they say. In scripture, a guidance fire led the lost by night. Similarly, every life experience guides and refines us, shapes our utmost potential. Cast in gold and cascading layers, *Pillar of Gold Fire* speaks to the beauty that each refinement could bring.

NINA VAN BART

My design practice is dedicated to increased sensorial experiences of space, shape and surface. With my designs of settings, objects and surfaces, the aim is to change the physical and emotional interactions between people and their surroundings, led by texture, colour and motion. I work in the field of art direction and set- and surface design.

The overall objective of my design practice is characterised by faithful attention to both the sensorial qualities of objects and their surfaces and the interactivity be-tween people and their settings. And so, my projects can be considered as relation-al: I challenge the ideas of how we usually connect with our environments and the objects they include – the aspects of a situation, the creation of a mood, the ar-rangements of our spaces and the rhythms of their elements, a different outlook and feel for surfaces, …. I reconsiders the original conditions and, with the outcome, intensifies our relationship with and the experience of our settings.

Two years ago I started a project Zooming In, Zooming Out, an ongoing material research into surfaces. This research results in an ongoing growing material library with several surface modifications. With Zooming In, Zooming Out I want to develop new material characteristics and a new perception of material via existing techniques, that offer solutions to technical demands such as acoustics and light as well as stimulate sensory experience of touch and adding spatial, visual effects to our spaces.

The first result of this project consists out of a series of four rugs. The series was developed in cooperation with the csignLab-team of Carpetsign. Recently Carpetsign developed a 3D tuft technique which I exploited within this project. This resulted in spacious, acoustic rugs with a very own identity which can be applied both on floor and wall.

The rug that I present at Venice Design 2019 is part of 'The Circle' exhibition of Dutch Invertuals, at Salone Del Mobile Milan 2019, and is the latest rug of my development.

DONALD BAUGH

Marking his 20th anniversary as a Designer/Maker, London based Donald Baugh celebrates by showcasing his 'Light Vessels' collection at the Venice Biennale 2019 exhibition.

After training at the noted Ryecotewood College in Oxford, United Kingdom, Baugh earned his honorary BA in Furniture Design at Middlesex University where he found his passion for sustainable materials and simple, solid design merging naturally and immediately in his unique bespoke pieces.

Over the years, Baugh has taken his intuitive approach to create custom interiors for both residential and commercial spaces, always mindful of the form and function of every piece and its longevity in its habitat. Alongside his commissions, Baugh continues to experiment with layering wooden and metal, always inspired by his passion for travel and the outdoor environment. His personal visual sketchbook of research in art history, architecture and interior design is always his reference point and core of his storytelling.

As an environmental activist, Donald Baugh collaborates with tree surgeons and the Forestry Commission to ensure that fallen indigenous trees like ash, cherry, beach and walnut are conserved appropriately. This has been a huge concern in the United Kingdom in recent years; therefore a timely response is important to maintaining the legacy of these special woods. Baugh enjoys celebrating their British heritage and practical functionality by incorporating these woods into his legacy luxury pieces.

Baugh's well-received exhibitions have been of his bespoke furniture in the past both in Europe and United States of America. For the first time, lighting will be showcasing in the form of a unique range of vessels, exclusively for the Biennale. In the spirit of design with intention, light seemed a natural progression as a sensuous addition to Baugh's furniture. A symbolic statement of his evolution as a designer as well as a sculptural instillation to stand on its own true merit for and in the prestigious Venice Design.

BC BIERMANN
CAVAD

As a digital urban artist and Associate Professor of Emerging Media (CAVAD) who is also a Southern California surfer + skateboarder, my life has been marked by tensions between old and new, high and low, inside and outside, analog and digital, private and public spaces. As a result of these binary collisions, my public works often reflect a kind of analog futurism, a design aesthetic that synthesizes these dualistic life experiences and creative impulses. Venice (Italy) and Venice (California) are old and new, high and low, artistically discrete but connected by distinct cultures that have been formed in and around the sea. One historically recognized as a birthplace of the Renaissance and the high art of Giovanni, Giorgione, and Tiziano, and the other a birthplace of distinctive surf + skateboard subcultures and the more urban, commercial works of Stecyk, Alva, and Peralta.

Highly nomadic as they move from break to break, often with skateboards in tow to kill time during flat sessions, surfers have created a unique mode of design through converting Volkswagen "busses" into residential spaces. Since the 1960's, typically cash-strapped surfers often resorted to living in their vans due to their mobile lifestyle and the high costs of Southern California. These vans are frequently a blend of iconographic imagery and pure functionality. Aesthetically, there is an eclectic mix of wood and metal surfaces, graphic logo stickers, posters, and surf + skate related paraphernalia. Functionally, there is often a bed, sink, custom lighting, stereo, and cooking equipment. In short, these vans offer the most undistorted window into this gritty subculture and arguably the only uniquely Californian design aesthetic.

The central arrangement of this interactive video installation is a van interior replica constructed as a VFX composite. The video beyond the back doors is a kind of Venice Beach cinéma-vérité shot on location. In augmented reality, tablets transport the viewer from the cramped interior of the van above the legendary Venice boardwalk, floating like a disembodied observer moving from inside to outside, from low to high. The bright lines that crisscross the walls are evocative of classic skateboard graphic color palettes and visually connect the installation to the organized chaos of the freeway networks that intersect and surround the beach cities central to Southern California design culture. Not pictured, but a part of the physical installation, are two companion videos: a curation of classic and contemporary surf + skate videos often found in local equipment shops and a data visualization of annual tides that exists as a connective maritime pulse of both seaside cities.

LORETA BILINSKAITE-MONIE

JOURNEY OF A THREAD

In the years of my youth, I experienced the wonderful simplicity of Lithuanian village life. I was immersed in a world steeped in tradition and taught to respect the heritage and customs passed down through the generations before me. As an adult, I made a choice to move to the UAE, which over nearly two decades, has become home. During these years, I have observed and admired the many similarities that appear in both cultures. My work on Journey of a Thread looks at the love of weaving that is common in Bedouin and Lithuanian cultures, both of which explore depths of folklore and demonstrate communication powers hidden within this craft.

Al Sadu, an ancient Bedouin tribal weaving art form, is rhythmically linked to poetry, memory, the weaving practice, and the graceful movement of a camel. Lithuanian weaving is commonly manifested in the sashes that form a part of the national costume, a hugely important piece bestowed on a person at birth and used for ceremonies and festivities thereafter. Both cultures communicate messages through their patterns and designs, telling stories and creating identity. Historically, this has been women's work and just as women bear children, they also carry forth many of the traditions through time-honoured practices such as these.

Journey of a Thread tells the story of threads that connect me to the two countries I call home. I have travelled and lived between each culture, adopting and absorbing many elements of both. Here in this work, I tell my story in the patterns and symbols I have chosen to use and have adapted the materials to illustrate my own unique journey. Each bead contains a wish or a prayer, and in the process of creating the pattern I share a part of my life experiences. The rhythmic movements, the sounds of the beads as they move between my fingers, sing songs of my travels and share stories of my journey. They remind me of morning prayers that wake me, and the sounds of the loom clacking back and forth in the adjacent room of my grandmother's house. The thread holds and supports me as I cross between two cultures, it ties me to my memories and to the things that have influenced me.

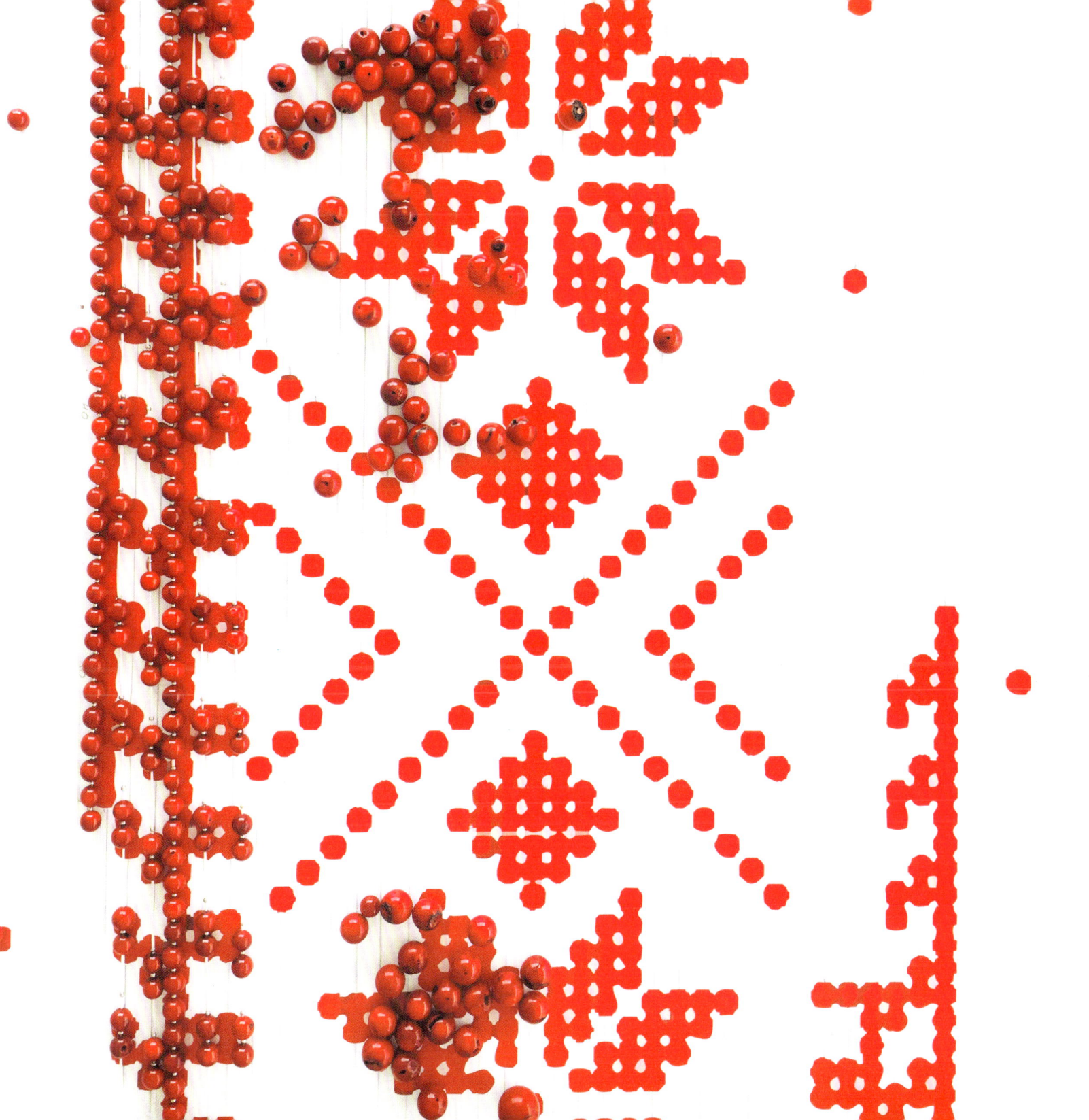

FELICIA BJÖRKLUND

LIGHT REFLECTOR

My work is the result of experimenting with shape, color, and light, and how this in an evocative way can interact in an architectural space. It is composed of a number of similar but not identical shapes made out of wood.

The bright color at the back of the form is reflected against the wall where a play of indirect light appears. The intensity of color changes depending on the character of light as well as on the angle from which you look at the work. When the hidden color is reflected on another area, what is invisible suddenly turns to something visible. In contrast to the wooden shape, the colored imaginary light is ethereal. I want to create a dynamic interaction between these two: the actual three-dimensional physical shape and the intangible light caught on flat surface.

The shapes can be repeated in diverse compositions, sizes, colors and numbers. To me, the light reflector is not only a piece of artwork; it is rather an idea, and an opportunity to develop a new design in a composition that will differ from the previous ones. Sometimes it might be just a slight change. Sometimes the most significant difference is how the light reflector appear when being installed in a new location and how it reacts with the light and the character of that building.

I am working as an artist and designer and make two and three dimensional art, site specific installations and public commissions. In my art I often work with senses as ambiguity and doubts and try to find a dynamic between surface and depth, between accuracy and diffuseness. I am very interested in how art interacts with the site's different spatiality, architecture and context.

DANIELA BUONVINO

RUSHES
UP-CYCLING&INTERACTION

Rushes is an interactive installation which translates sounds into light. It's created to remind you the sounds of nature and the sounds of humanity. It's created to bring you back on earth and let you feel your body in space.

You are here, now. The experience is based on simple elements which are universally recognised: sound and hearing. We have the power of interaction with all the surroundings and we should always remember our connection and magical relationship with the nature. This installation is 90%made by reused material. Nowadays we consume too much and we save too less. The reeds are made by reused beer barrel and metal tubes and the leaves are made reusing homemade Granny's cotton crochet, an old memory of my family, made with the antiques knitting technique. The beer kegs are an elegant industrial packaging which is used for only a week and it's afterwords ending his short life in a garbage bin. Recreate reeds with plastic barrel is a sign of respect for nature. Don't waste, but create. No body is taking time and energy tearing apart products made by different materials which means, as we all know, this products are not been recycled. They are potentially recyclable, but no body does it. It's easy to throw away staff, forgetting to think from a different point of view. We should stop buy new objects every single day and get back to creativity: use what we already have and learn how to reuse object in a different way. Rushes inspires visitors, thanks to their creativity and natural movements, to go back to a more natural life.

Different visitors will create different lights experiences and unpredictable scenarios.

As in life we have the power to decide and to modify the reality around. Creating a light experience is a way to underline the connection between elements and human being, between colours, sounds and our movement for example, one of the connection is that all of this three element are producing waves of frequency. About the artist: I am an Italian Industrial Designer, based in Amsterdam since 2009. In 2015 I finish my master in Industrial Design, at KABK, Royal Academy of art in Den Haag. Fascinated by metamorphoses and behaviours my works swing between Light Designa and interactive art. I produce the lamp my self, upcycling plastic beer barrel used in pubs for tap beer.

After one year of experiments and small exposition I was looking for different craft methods to give a personality to my products, and in 2015 I start to collaborate with programmer and electronic engineer Walter Luppino, light and a really creative metalworker Stefano Nocciolo. Thanks to this collaboration technology meet art; power led, sensors and motherboards are hided inside unique, manufactured lamps. The interaction with the public is a challenge to create an experience, surprising, sustainable and memorable.

KEVIN CALLAGHAN
IN COLLABORATION WITH A2 ARCHITECTS & STUDIO MEREDITH

PRECARIOUS GROUND

This artist/architect proposal begins by drawing inference to the delicate balance that Venice repeatedly lays bare - our ever finer relationship with our built and natural environment due to rising sea levels.

This proposal begins with a vunerable construction of ground. The City of Venice is a completed constructed artefact founded on estuarine marshland in a low-lying lagoon. Built atop long wooden piles 60' feet in length and driven deep into the ground, these oak and larch piles go deep down reaching past the weak silt and dirt to a portion of the ground that is hard clay which can then hold the weight of the buildings above. The waters of the lagoon carry an extremely large amount of silt and soil and the wood has been blasted by this sediment for years. The wood has absorbed the sediment and has quickly petrified into basically stone at an accelerated pace.

Originally the piles were placed as closely together as the soil of the ground would permit. Stone and rock were placed in between the piles which kept the silt from rising up during subsequent pilings. Two layers of wood was added on top of the piles which is where the masonry starts. A marble damp-proof layer was used in the masonry build-up due to its impermeable to water.

Poignant to this collaboration is the reciprocal relationship that Venice has with the Irish city of Cork – one of the many Venice's of the North – which faces the very same predicament of piled construction and rising sea levels. In the instance of Cork the present solution is to build crude flood defence walls at the heart of the city that destroy any meaningful relationship to the history of built edges to date. It is the intention of this installation that both cities will be brought into critical dialogue in this proposal through a series of islands made from clumps of piles that support fictional city fragments made of fired clay.

Each city fragment atop each clump of piles explores a number of constants in each city: the relationship of solid to void; the relationship of ground datum to roof profile; Two colours of clay are used in the making of the city fragments – a natural grey clay similar to the silted clay of the Venice lagoon and a red ochre clay that points to the pigmented plaster and terrazzo floors found throughout Venice. The wooden plies are stained a grey-black that points to the petrified piles of Venice.

Positioned within a canal-side first floor room these city fragments will be seen at eye height and will be further transformed in the space of the room through the reflections cast from the Grand Canal beyond.

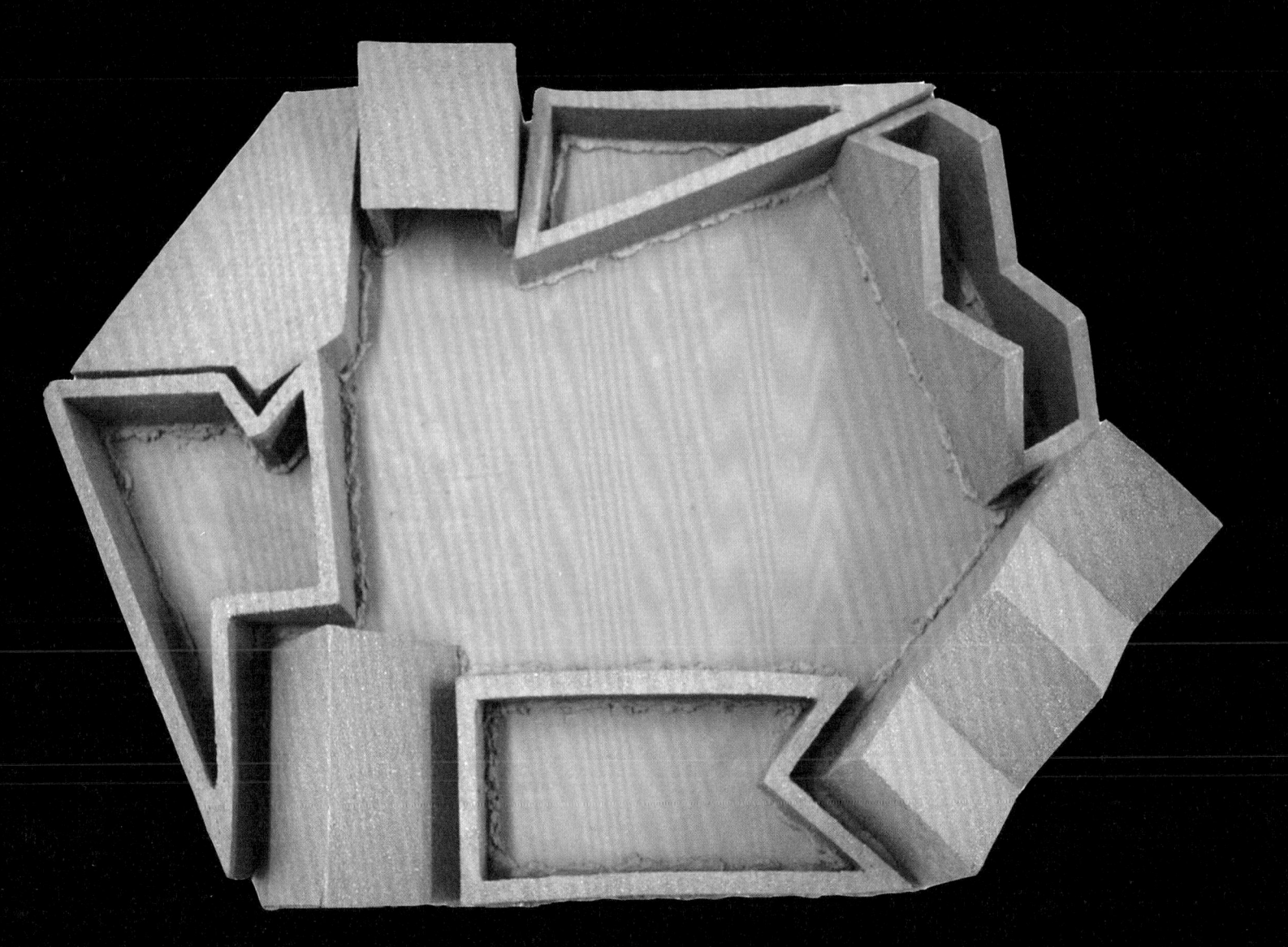

BARBARA CALVO DESIGN

I like designing objects marked by simple and clean lines, at the same time revealing vigorous and imposing qualities. Big sizes and massive structures give style and personality to plain geometrical shapes, that's what makes my creations key elements of the space they inhabit. I love metals because "they can be all or nothing", they are exceptionally various and changing: poor or rich, soft or tough, rough of refined, cheap or expensive, shiny or opaque… in any case, authentic and natural. Because metals are the sons of our Earth!

An object can be given a totally different appearance and personality depending on the choice of details and finishings. A visible weld joint, an unrefined metal sheet, a rust stain, they can all surprisingly add style and elegance, even in a classic ambiance, by revealing the magic of contrast and of "provocation"! I chose metal for the greatest par of my collection because it reveals my way of being and my way of interpreting interior design.

My research aims at being minimal and warm at the same time. Iron has a rigid and unbending nature, it's almost austere, but it's extremely receptive if exposed to heat… With fire you can shape iron and melt it, dominate it and frame it to your will. In a moment it turns from rigid to fluid, achieving unexpected and surprising effects! Each of the objects I've created, though made out of metal, hides a burning soul, a soul of fire and passion - the same passion that moves the energy for a true and great life.

PIERO CASTIGLIONI

1954

The new lamp developed by Barrisol® is inspired by the X Triennale industrial design exhibition wich took place in Milan in 1954, 60 years ago. My uncles Achille e Piergiacomo Castiglioni, in collaboration with the architects Roberto Menghi, Marcello Nizzoli, Roberto Rosselli and Augusto Morello (curator), Michele Provinciali (graphist), Lorenzo Pepe (sculptor) and Mario Reggiani (pinter), created this exhibition's laying out.

1954 was a very important year for Architecture and Industrial Design with the first International Congress organized.

The specific lighting system created for this exhibition including the creation of huge canvas bells was my source of inspiration for the design of this new Barrisol® lamp, as a tribute to the Castiglioni brothers but not only. They were used at this time to create a huge "lighting ceiling", to diffuse light without apparent shadow.

This caracteristic is now possible thanks to the Barrisol® system which combines the technicity of a special metallic circle (aviable in 4 dimensions 80-120-160 or 200 cm) and high qualities of the Barrisol®, Blanc Venus membrane offering a perfect light diffusion with more than 0.5 light transmission coefficient.

This new material replaces perfectly the original "coquille d'oeuf" color sheet, thanks to the integration of new generation LED.

1954 lamps offer an excellent light transmission and a perfect color rendering guaranteeing a remarkable economy of energy.

The 1954 lamp can be installed in all interior spaces thanks to the 4 dimensions available.

SOPHIA CHRAÏBI GIORGI

THE PRAGMATIC DESIGN CONCEPT

Characterized by a mono apparent material, all furniture from SCG manifest a minimalist bias. Only consists of modules that fit in recess, without the use of hardware, each element plays a structural role. And the absence of ornament joined the simplicity of constructive geometry, where only the function defines the shape. SCG pragmatic eco-design furniture is fonctional, modular and sustainable, reated with HDF, or high-density fiberboard, a water and fire resistant engineered wood product.

The digitial cut process to manufacture wood structures allows greater precision. Inspired by the sincerity of the material productions Arts & Crafts, Bauhaus minimalism and Mingei philosophy, the pragmatic design furniture SCG is the contemporary expression of the increasing mobility of people and their desire for uniqueness.

GENESIS, THE ROMAN INSPIRATION

The domus, or Roman homes, discovered in Volubilis revealed the great legacy of the Roman Empire in North Africa and a typically Mediterranean architectural layout. Homes were organised around a central open space. Opening onto this courtyard (atrium) were the primary living spaces, including the triclinium : roman dining room and a sitting area for receiving guests, with three chaise longues. The Ksour, or castles, of Morocco's Atlas Mountains made use of a layout similar to that of the Roman domus: a central courtyard with a dedicated sitting area for receiving guests.

This Berber lounge, referred to as a tamesrit, featured a series of alektu, or backless sofas, whose etymology can be traced back to the Latin lectu, or bed. Similarly, the Latin verb sedeo/sedere, meaning either "to sit" or designating the seat itself, suggests the Moroccan Arabic term seddari. Still in use today, these seddari are what we often refer to as a traditional Moroccan lounge. Even the riads built later have taken inspiration from this typical Roman architectural layout, creating a sitting area lined with seddari near the entryway that opens onto a central courtyard.

Following in the footsteps of the Roman triclinium and the Berber tamesrit, the seddari is as important today as ever.

IN TRIBUTE TO ROMAN LIFE
AROUND MEDITERRANEAN SEA

Papyrus is a removable daybed esthetically inspired by the Antic Roman way of life, created for a luminous patio of a contemporary ryad, in tribute to the great legacy of Romans in Morocco.

Its modular blue framework is composed of 3 kinds of elements that fit together without any screw according to the eco conscious philosophy. For the daybed *Papyrus*, the specific form of the mattress accompanies the armrest and is trimmed with pillows; its ecru upholstery is a handmade weaving wool fabric, setted off by golden artisanal decorative elements, usually designed for traditional dresses. This ecru/gold moodboard is an other reference to the legacy of the Antic way of life. Those feminine curves invite to enjoy laziness and another perception of time.

CLOUDANDCO
YEONGKYU YOO

I was stunned when I saw a giant build-up of plastic in the sea from BBC, and it made me concerned about serious effects out of that. As an industrial designer, I wanted to make a good impact on it with design value. Since consuming plastic bottles would be one of the biggest sources of plastic waste, providing safe water to drink could reduce the plastic waste drastically. This brought the idea to me cooperating with Coway, a water purifier maker in South Korea.

Every minute, 1,000,000 bottles are bought in UK. However, less than 50 % is collected for recycling, yet only 7% turned into new bottles. We all know that consuming plastic bottles makes terrible problems to our environment, but it is hard not to consume them due to its convenience. In order to encourage people not to consume bottled waters, alternative way to drink clean and safe water needs to be proposed.

Many cities have been working on providing free water in public, installing water fountains, so that people could consume less plastic bottles. However, accessing to free water in public won't be enough because the concerns about the cleanliness of public water, fountains and dispensers still remain. Although the tap water was clean and safe at first, it could be contaminated while traveling due to old pipes or other matters. These concerns make people to drink bottled water. Therefore, I designed Water(refill) Station in the city with Coway, new generation of water fountains with purifying technologies. Water(refill) Station offers safe and clean tap water showing the real-time cleanness with the gauge window and mobile App. It's cost-effective and environmentally friendly. People can check how the maintenance has been done and its cleanness with real-time water monitoring system. It makes refilling the water bottle easy, and the mobile App allows people to search a nearest Water(refill) Station from where they were. In addition to all of this, it is a beautiful object in the cities.

The focus of designing Water(refill) Station was making the refill of a water bottle easy so that people can carry their bottles and stay hydrated, healthier lifestyle. Conventional water fountains were more for taking a sip of water with the sink, little tricky for filling the bottle. Also, unnecessary sink is eliminated since it can be germ hotbeds. Its prototype to realize the design concept is made by an exceptional model maker in Seoul, Model Solution.

To motivate people, the reusable bottle is also designed made of biodegradable plastic. People carrying the water bottle will attract others' attention like a campaign booster. If 8,000 liter of drinking water from Water(refill) Station were consumed, it would prevent from using 16,000 disposable plastic bottles. This will raise public awareness helping reduce the consumption of plastic bottles gradually.

Water(refill) station will offer cost-effective and environmentally friendly way to drink healthy water as well as making our cities more beautiful.

LUCE COUILLET

Luce Couillet uses weaving techniques to create one-of-a-kind and limited-edition mobiles.

Articulated around a hand-woven "spine" and decorated with different materials (wood, horsehair, paper, metal), weaving acts as a binding and assembly vector to bring the mobiles to life.

The central axis creates a stable, guiding structure which anchors the different shapes revolving around it and moving according to the artist's intention. The overall effect of the materials and techniques used is an authentic and subtly contemporary mix.

Luce's finishing touch in this artisanal approach to construction is the use of laser technology, which she employs to cut the mobiles' exacting edges before or after weaving.

The shapes obtained are simple, seemingly unsophisticated. But this refined geometry is revisited through transparency and anamorphosis.

By playing with different scales, the mobiles become decorative elements or real architectural drawings.

Luce works on both personal designs and custom orders.

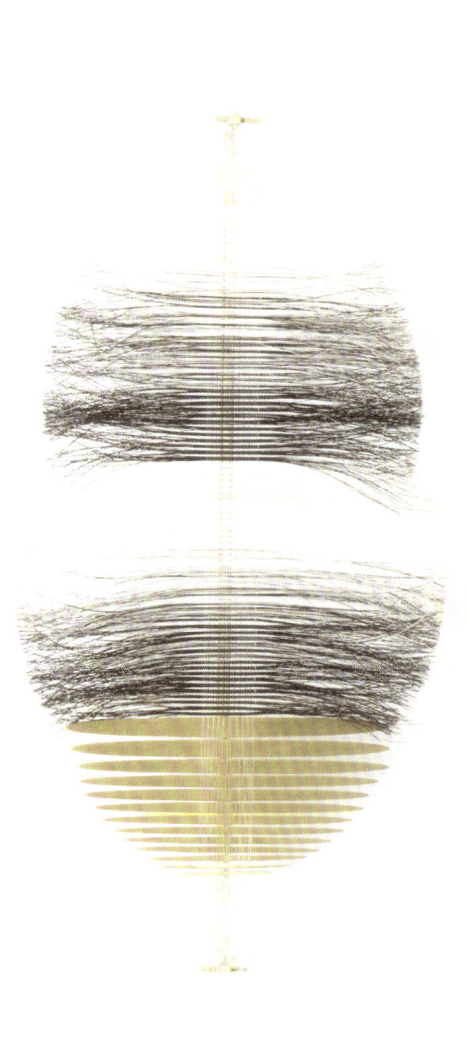

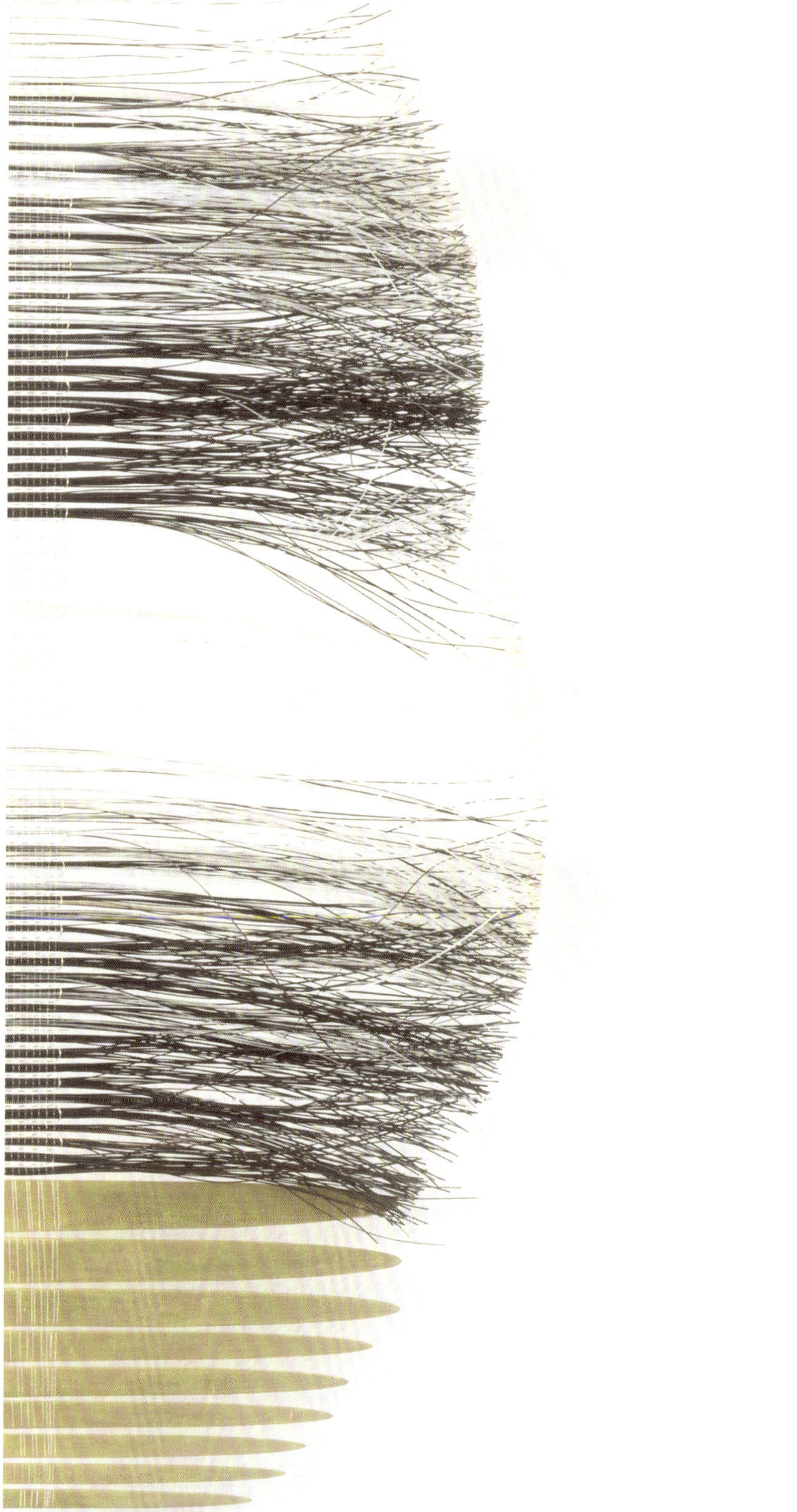

CREATIVE CHEF
JASPER UDINK TEN CATE

Creative Chef Studio Founder Jasper Udink Ten Cate and the studio's concept developer Cisco Schepens constantly investigate new ways of designing. This spring they went into the music studio and used sound data to design products. A cutting-edge new project arose. The Composition Table, an interactive experience in which the dinner guest becomes part of a musical composition. Creative Chef invites guests to use ceramics, sections of tablecloth, napkins and cutlery to interact with an artificial intelligence system. While exploring the possibilities of the tableware, a personalised musical composition comes to life. With the Composition Table Creative Chef redefines the typology of our eating habits.

The Composition Table is an entirely new approach of food, changing the eating experience from a passive consumption to an active participation in a live musical composition. We add auditory perception to physical tableware design. In product design, as in life, there is a strong tendency to follow acquired patterns. Learned behaviour is repeated ad infinitum. Creative Chef breaks through this cycle by asking what-if questions. What if... we simply don't follow the known? What if we start designing objects in a music studio instead of an art studio? Creative Chef examined the possibilities of utilising the wonderful shapes of sound vibrations for product design. Instead of sitting down and drawing, we started recording and exploring the phenomenal world of sound and vibration. We wrote our own music and captured the physical data. This data was used for the visual designs of our new tableware series from tablecloths to cutlery and napkins to glasses.

These objects are the starting point of a state-of-the-art food experience. An experience to tickle all senses. The Composition Table is composed of a wide variety of interactive products, each with a unique sound. Through the existence of the table you are given an opportunity: to explore the sounds behind the objects, engage the AI system by scanning the tableware, and to re-view learned patterns in design + food. By adding more objects, the auditory experience changes. You are the composer of this food performance. Creative Chef adds digital content to analog products with the technical support of Dutch tech company Superp. Superp invented the ingenious AI system behind The Composition Table. The project serves you a new way of storytelling. A musical journey and engaging method of communication with dinner guests. Music design becomes interactive product design. A new paradigm in Design Thinking; an innovative way of adding content to products. We believe that the true power of design is developing stories which will be integrated in peoples' lives. Experiencing sophisticated design enhances these stories and creates precious memories.

DÁ DESIGN STUDIO

As a very diverse people, Nigerians and Lagosians in particular, are constantly in search of visual and graphic representations of our collective identity. Every now and then we hit a gold mine and milk it for all it's got. The vivacious Danfo buses, the good, bad and ugly of public transportation in Lagos with their unmistakable yellow and black identity are one of these gold mines. Their notoriety has existed simultaneously with only a few other things as identifiers of Lagos itself. Even with much controversy surrounding Danfo buses, we have continued to identify with the strength of the Danfo aesthetics because just as any group of culturally linked people, we crave a collective image. As a studio, our design quest was to find another way to connect with this identity outside of the common yellow and black direction and all its siblings. With the constant threat from the government to phase out Danfo buses, it will be a shame for them to go without the unique typography they shared being robustly recorded from a visual standpoint. The point isn't whether it's good or bad, fine or ugly rather the point is that it is worth understanding & recording. The font Danfo Std provides another answer to the question of identity and participates in robustly recording Danfo visual culture. If we're going to keep milking the Danfo identity, the milk is better richer.

SAVERIO D'ELIA

LUNENOTT

WHY A NEW LAMP?
The design of a new lamp is justified / dictated by the desire to reconsider the relationship between man and material, from the need to rediscover their own space without being gripped and suffocated by piles of trivial things, by the redefinition of the idea of a misunderstood accumulation. It is a return to substance, simplicity of form, beauty combined with essentiality.

WHY THIS FORM:
Lunenott, with its geometric, archetypal shape that finds "its raison to exist within an unconscious and hereditary knowledge intact through our biological memory" (Jung), releases an ancestral energy.

WHY THIS COLOR:
It is the perfect shape, which Kandinsky associates with the blue color, producing a perpetual movement that represents the stars, it becomes a small universe where every form of passion is condemned in favor of the rigorous coldness of the method.

WHY THIS CUT:
The presence of an incision and an absence of the material lets the emptiness behind emerge, evokes an opening towards the elsewhere, an escape route, but in Lunenott the cut opens the light in the dark and underlines its presence.

DUODU
RITA NYLANDER AND ANNE GRUT SØRUM

DUODU consists of the knitting designers/craftsman Anne Grut Sørum and Rita Nylander. They have worked together since 1998 with a common anchor point in the joy of knitting, humour and curiosity. A curiosity that has brought about experimentation and new techniques and expressions during the entire period. They are found in the intersection between art, craftsman design and design. DUODU is not changeable fashion, but timeless elegance, blended with a neat touch and humour. They design according to the motto:
ENJOYABLE - SIMPLE - ELEGANT.

Even though DUODU is not preoccupied with fashion as such, we find traces of the trends of our time in the group's work. The inspiration is complex and moves effortlessly between various geographies, times and themes. They collect just as much from literature, music and film as from their local community and the everyday life they are surrounded by. They seldom make use of the obviously grandiose, but will rather focus on details, investigate structures in the wall, reflections of light in melting ice or the colours and patterns of insects and animals. The impressions are transformed through a long process in which the starting point proper is not always easy to wind yourself back to.

The exhibition "omHAVNelser" collects impulses from the harbour in Trondheim where they have their studio. For a number of years, Rita and Anne have photographed the everyday life outside their studio and become inspired by the changeable rustic life of the port, a cultural heritage that is changing rapidly. The area is still filled with the activities of the port, but the city is steadily moving closer. The previously unkempt area is in the process of having a makeover. Anne and Rita have tried to capture what is about to get lost.

DUODU wants to make creations for eternity and the most beautiful has not been created yet.

EDINBURGH COLLEGE OF ART
UNIVERSITY OF EDINBURGH, MAL BURKINSHAW (MA RCA) WITH SOPHIE HALLETTE LACE

'SILHOUETTES EN DENTELLE' 2013-2017

Fashion is a powerful medium, conveying influential social messages concerning beauty and body image. As an activist for improvement around themes of diversity and well-being through fashion education, I seek to explore how to utilise fashion design to create reflective debate concerning themes of body image and beauty.

The design series 'Silhouettes en Dentelle' began in 2013 as part of the exhibition 'Beauty by Design: Fashioning the Renaissance' at the Scottish National Portrait Gallery, National Galleries of Scotland. The exhibition united art historians, curators and designers to investigate how we can use Renaissance paintings to question present-day assumptions about beauty and body image. The project investigated the narrative between Renaissance portraiture and contemporary fashion design, exploring how body image and beauty ideals remain in a constant state of flux.

The sequence of jackets were produced in response to the body shapes and garments of sitters depicted in the portraits of the Scottish National Portrait Gallery's Reformation to Revolution Gallery. The concept was to showcase Renaissance silhouettes 'painted with lace' onto contemporary tailored jackets. The series aims to express the identities, status and body proportions of historically important figures including Mary Queen of Scots, Lady Agnes Douglas; Countess of Argyll, Lady Arabella Stuart, James VI and I and Margaret Graham; Lady Napier.

The lace is sponsored by Sophie Hallette, based in Caudry France, manufacturers of the highest quality lace for global fashion houses. The choice to use lace directly related to the centuries of craftsmanship involved in making this delicate material, which was a signifier of wealth, status and hierarchy in the Renaissance. Black lace was favoured, yet was rarely seen in Renaissance portraiture of the time. The process of hand stitched applique took over 1000 hours and consists of 30 separate lace motif designs.

The series of jackets offers the viewer a contemplative experience to consider past and present notions of 'normalized' body shapes and beauty ideals. The jackets were designed in sequential scale, and do not conform to any specific size measurements. This poses considerations around age, body size and beauty; all highly topical in the current debate concerning fashion and well-being. The jacket design is engineered to be non-gender specific, and the juxta-position of lace (which is traditionally feminized) allows the audience to consider concepts of beauty relating to fashion and gender codes.

Mal is an alumni of Edinburgh College of Art and the Royal College of Art, and is Programme Director for Fashion at The University of Edinburgh. He leads the Edinburgh College of Art Diversity Network, uniting academics, students, charities and the fashion industry to explore how fashion education can contribute to improvements in consumer self-esteem through 'emotionally considerate design'.

FAINA COLLECTION

Live Ukrainian design, FAINA Collection, started life in 2014 as an artistic reflection on the Revolution of Dignity in Kyiv, Ukraine and desire of the whole nation for self-identification. Founded by architect and designer Victoriya Yakusha today design brand has evolved into a progressive furniture collection that celebrates Ukrainian design traditions & culture while embracing modernity.

FAINA Collection is based on Victoriya`s recent study of the domestic traditions, forms, materials and craft techniques, that were carefully transformed into contemporary minimalist design objects. Clay, felt, willow, flax, and solid wood are the main elements of the FAINA collection, helping Yakusha to express the concept of "live design" – design that combines the energy of eco materials and the simplicity of ancestor's lifestyle. The whole collection is designed to involve simultaneously all senses and encourages feeling the soul of objects.

For VENICE DESIGN 2019 FAINA is bringing the iconic objects that have become a hallmark of the collection: cabinet SOLOD with ceramic façade and set of vases KUMANTSY. Cabinet SOLOD ("solod" in Ukrainian means cereal grains) square shape wooden bar cabinet with a handmade clay facade with nerve holes all over, brings a feeling of healing energy to your home. Set of vases KUMANTSY ("kumanec" in Ukrainian - ceramic figured vessel) - traditional festive pottery with a hole-bagel in the middle, transformed into modern interior decoration, that has a power to unite people around.

BACK TO THE ROOTS —
ROLE OF CLAY IN FAINA COLLECTION

As an artisan and design tool clay is very plastic and timid, it interacts with the artist, influencing the character of his future piece of work.

Our ancestors believed that ceramic has a soul and can heal people, filling their hearts with warmth and their bodies with living energy. Small clay details at your interior will deliver positive vibes. While working on FAINA capsule clay collection we collected numerous samples of original Ukrainian pottery traditions and collaborated with several artisans, who still use ancient manual techniques. In their caring hands the seemingly fragile nature of clay turned out to be very reliable and looking modern in the aesthetic sense.

Victoriya Yakusha: "As a contemporary artist and designer I see my mission in making Ukrainian identity understandable and recognizable throughout the world. In FAINA Collection you will feel the force of energy and harmony between humans and nature that has been encapsulated over our lands for centuries. I stand for honest, sustainable and live approach to design. We are living in global world, surrounded with trends of individualism and minimalism, but in a "marathon" for trends we should value design that remains to live with us in future. Only this way we can prevent our national identity from vanishing".

HICHAM GHANDOUR

First and foremost, I am inspired by contradictions. The juxtaposition of seemingly opposing materials defines my work. With these appliques, I have chosen to bring together materials that can give a dramatic contrast: rough and smooth, dark and light, opaque and translucent. Made of casted bronze and rock crystals, these pieces come together to create something that is unexpected and unique.

Furthermore, I find inspiration in my travels across the world, from places that widen my horizons and spark my imagination. With these appliques, I personally selected and imported the rock crystals. By using rock crystals that are almost transparent in their natural form, I have created a piece that allows the light to glow through the semi-precious stones.

As a gilder by trade, I am committed to being involved in every aspect of the production of my designs. Here, I worked with artisans through every step of the creation and production. I molded the bronze frames by hand, working directly with the foundry. This allowed me to create a final product that is reflective of my philosophy and my designs.

These appliques, as is the case with all my designs, are imagined to be timeless. They are not reflections of our time, but rather designed to be used and be useful beyond any restrictions of time and space.

Designed as a pair, these light appliques are a contradiction of materials, a unification of worlds, and an experiment in blurring the lines between design and production.

GHYCZY
BY PETER GHYCZY

"For me design is an obsession. It's the ongoing quest for new innovations in the construction of an object. I design the same thing over and over again until I have found the optimal solution – an object with a simple aesthetic and minimal use of materials, that is durable and easy to assemble. To me good design doesn't have to be complex or opulent – its appeal lies in a timeless quality that will last for generations."

Peter Ghyczy

MADE TO LAST

GHYCZY was founded in 1971 by acclaimed architect and designer Peter Ghyczy. From the start Peter Ghyczy was passionate about timeless, sustainable design. This is today known as emotionally sustainable design. His products are created with great care and made to last. The natural materials GHYCZY uses are carefully sourced. Their surfaces are finished in a such a way that time will only enhance their beauty. All furniture is made by hand in the studio in the Netherlands.

The Garden Egg Chair, designed in 1968, and the light weight chair S02, launched in 1986, are two of Ghyczy's most iconic pieces. The complete collection now comprises tables, seating, shelving, cabinets and lighting; only made to order and hand-crafted using sustainable materials with high quality patinas – to be enjoyed all day, every day.

Today GHYCZY is still a small boutique family-run company. In 2001 youngest son Felix Ghyczy joined the company and has been working side-by-side with his father to develop an iconic collection appealing to all generations.

GHYCZY furniture shines in both modern and classical interiors, in private homes, luxury hotels and museums alike; GHYCZY can be found in the Vitra Design Museum in Weil am Rhein, the Victoria & Albert Museum in London, Christie's auction houses, the Ace Hotel and the Ritz Carlton.

URBAN AUDREY GP05 / 2016

The frame of the armchair is made of round tubes and solid wooden rods. The chair is lightweight. The backrest and seat are loose and can be adjusted to meet your individual seating preference. The cushions are soft and comfortable. The craftsmanship combines metal and wood. They are elegantly connected in a visible joinery. This construction lays the base for a new series of functional furniture items named "SAFARY" by the design Peter Ghyczy. The chair creates a balance between craftsmanship and industrialization while using metal tubing and wooden.

URBAN JODIE 302 / 1988

The frame is made of one single round tube bent into its shape. Two sand casted joins are used to create the hinge for the backrest. The use of minimal material makes it a lightweight chair. The triangle construction guarantees stringent, rigidity and longevity. High comfort is ensured by the ergonomic and flexible backrest, which always supports ones lower back and makes the chair surprisingly comfortable.

DONNA GLUBO-SCHWARTZ

My work through Spatial Element has been a systematic consideration of expanding the boundaries of fiber. Having received my BFA at the School of the Art Institute of Chicago, where studying all mediums is encouraged. This integration of sculpture, color theory, and fiber arts, have invariably informed my work.

The process of working with and studying fiber as a medium and as a commercial commodity, is an exploration into human history.

Fabric construction has been attached to a form of feminine work and expression. The methods of transforming the surface of fabric has been a means of cultural and spiritual dialogue. Japanese resist dye, Shibori, is an ancient form of surface manipulation used with the purpose to decorate fabrics through color and design.

Traditional Shibori uses regulated sizes of fabric. I abandoned that formality and went big. That one step, lead to a new way of concidoring how to use this material. The silks I began working with, which is traditionally flattened, were pleated in its raw state, thus initiating my attraction and focus on three dimensional forms. The challenge was how to best show and feature the sculptural surface. Through a home project of laminating my flat fabric in glass for a door insert, the idea was hatched to capture the pleated fabric in glass. That resulted in a collaboration with a glass company to laminate the fabric.

My current interpretation is the Twisted Object. This newest work is focused on the architectural exploration of three dimensional formed materials.

All *Twisted Objects* are a transformation of fabric creating organic and ethereal shapes. Each configuration is a frozen moment suggesting anticipated change. Both form and color gives every Twisted Object its own way of relating to our experience of spatial acuity.

My current installation is a study in the color Fuchsia, with a range of saturations. The psychology of color informs my direction when choosing a specific color to work with. Fuchsia, also known as magenta, has been described as a color that brings harmony and balance in life. It is also associated with attributes of non-conformity and spontaneity. Black *Twisted Objects* are interspersed in the grid of Fuchsia. The Black gives weight in a show of strength, elegance and authority. A grounding element to the vibrancy of the Fuchsia. The intention of each Twisted Object is its insistence of individuality and change. Although fixed in a particular form, the possibility of alteration is real and achievable.

Thank you to the European Cultural Centre and the entire VENICE DESIGN Team for their assistance and support. Thanks to my family for their loving support.

JULIE HELLES ERIKSEN

Julie Helles Eriksen is a multidisciplinary designer from Denmark. As a designer Julie focuses on conceptual work at the intersection of identity, textile objects and technology. The work "Who are you?" challenges our self-representation of today and shows new possibilities in combining textiles with technology.

WHO ARE YOU?

We communicate to others through what we wear, what we post on social media and what we choose to hide. Online, we keep certain pictures and accounts private so that not everyone can see every facet of our lives with just a scroll or a click. We curate ourselves to present different versions: one for work, one for friends, one for family, and maybe one for who we dream to be.

Our physical lives and our digital lives feel separate. This garment merges the physical personality with digital personalities to create a more complete representation of the wearer, bringing together our different identities on the body to reveal a more versatile communication of the individual.

Our camera phone is the key to creating these online personalities. We take pictures of everyone, everything and every version of ourselves. The garment utilizes the camera phone as a means to unlock the various personalities of the wearer of the garment. The woven QR codes can be scanned and they link to the individual's social media accounts and other online identities. As the wearer moves, the garment unfolds, revealing new QR codes and allowing different information to be accessed, and a more complex identity of the wearer to appear.

The inside of the garment exhibits a mixture of private and public images, representing the wearer of the garment.

WHAT'S E·SPORTS?

E-SPORTS

[e-spɔːts]

Perhaps E-Sports is vague, premature and even paradoxical. The topic of E-Sports Stadium is a new chapter for architectural exploration, and must be continuously developed and interpreted....

E-Sports Spirit
Electronic sports (E-Sports) is a sport which players compete with each other through playing video games. The influence of E-Sports is spreading tremendously around the globe. As an evolving type of sports, E-Sports may soon become an official Olympic Sport in 2020. Without any limit to race and boundary, the introduction of E-Sports allows us to build collective effort for pursuing human excellence. Advanced technology could also bring us to a whole new perspective on the evolution of sports, which greatly facilitate electronic digital influences.

Fluidity
The Design of the E-Sports Stadium adopts the approach of Indoor Stadium, which embraces the idea of designing the spaces to be dynamic, floating, and ever-changing space.

Imbalanced Tension
The Design of the E-Sports Stadium is based on Asymmetric Symmetry. E-Sports dynamics is based on the foundation of two parties' tournaments. The Synergy and Energy involved in the tournaments suggested an Imbalanced motion, which attains a sense of equilibrium similar to the concept of "Yin and Yang". The LED TV on the canopies and the curtain wall could provide the audience with much better visual effect even from the outside of the Stadium. Public who are outside of the stadium could also feel the excitement of the ongoing tournament.

Flying Keyboards
The E-Sports Stadium is designed to be covered by a pair of "Flying Keyboards", which purposely projects an asymmetrical beauty. The Design of "Flying Keyboards" symbolizes two parties in competition, revealing the competitiveness and synergy beyond the tournaments. The Uprising and Uplifting form of "Flying Keyboards" attains juxtaposition thus well redefining the Imbalanced Tension in harmony. The asymmetrical "Flying Keyboards" reveals the Chinese philosophical concept of "Yin and Yang". The concept of "Yin and Yang" suggests that two opposing and contrary forces may actually be complementary, interconnected and interdependent to achieve equilibrium between the competitors in E-Sports tournaments.

IS E·SPORTS A SPORT?

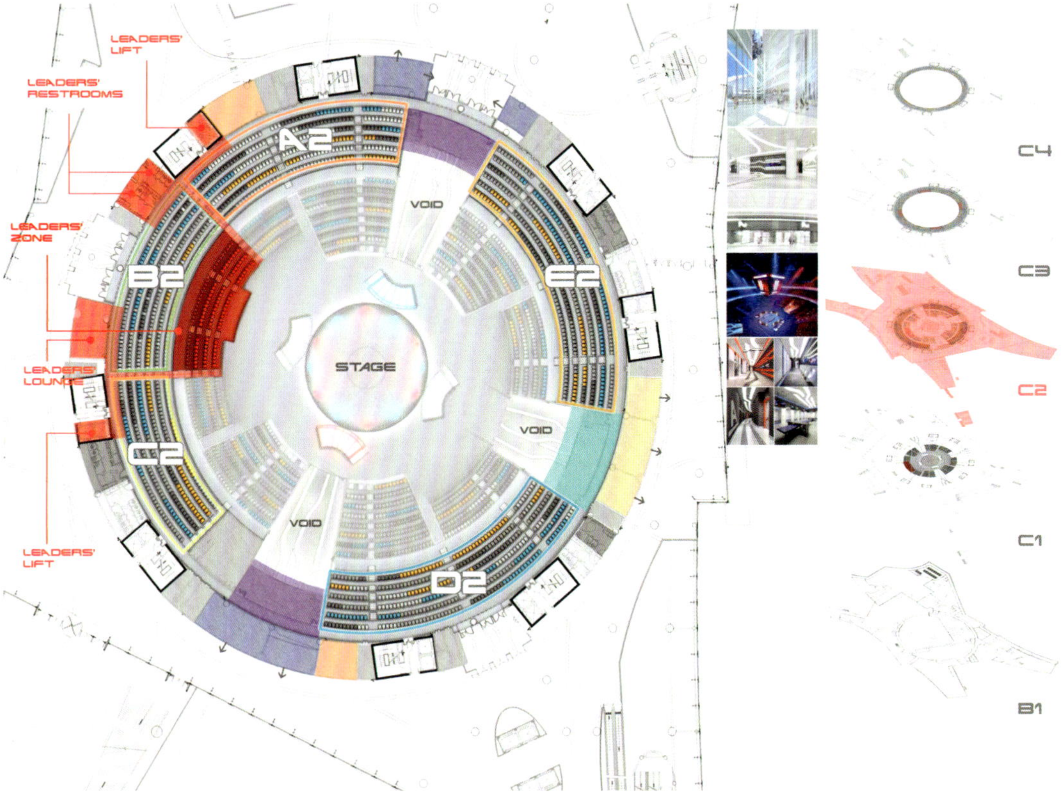

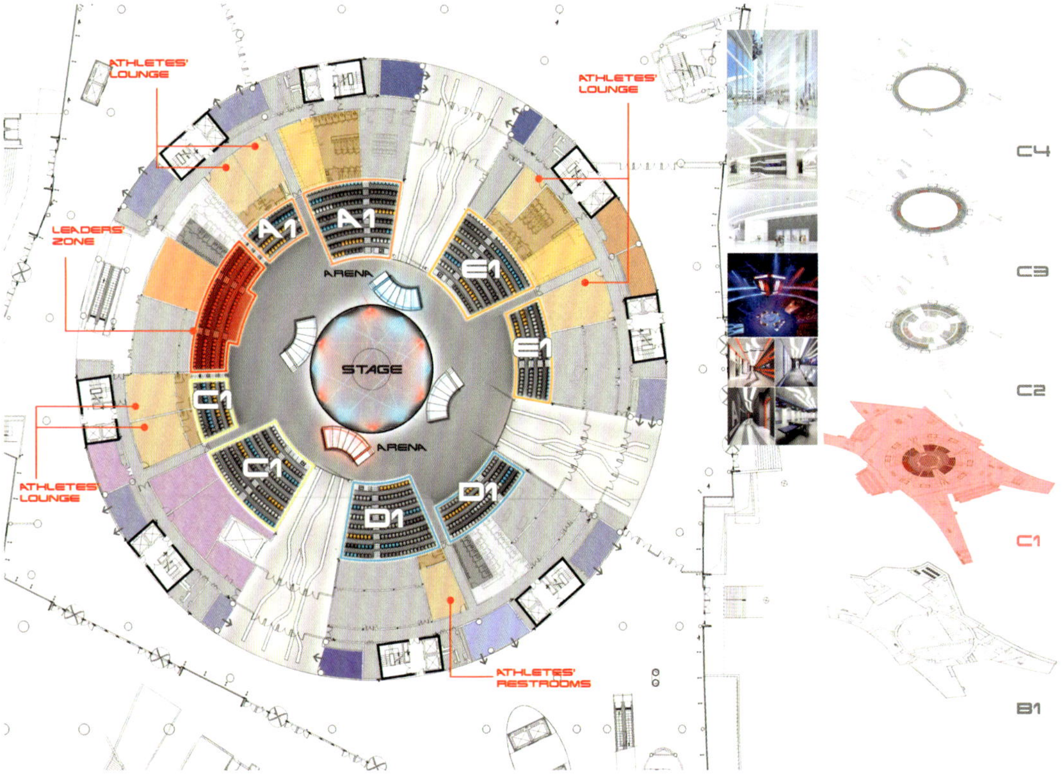

INTEGRATED

E-SPORTS
[e-sports]
...can be applied to co
online multiplayer role pla
games, as long as there i

West Elevation

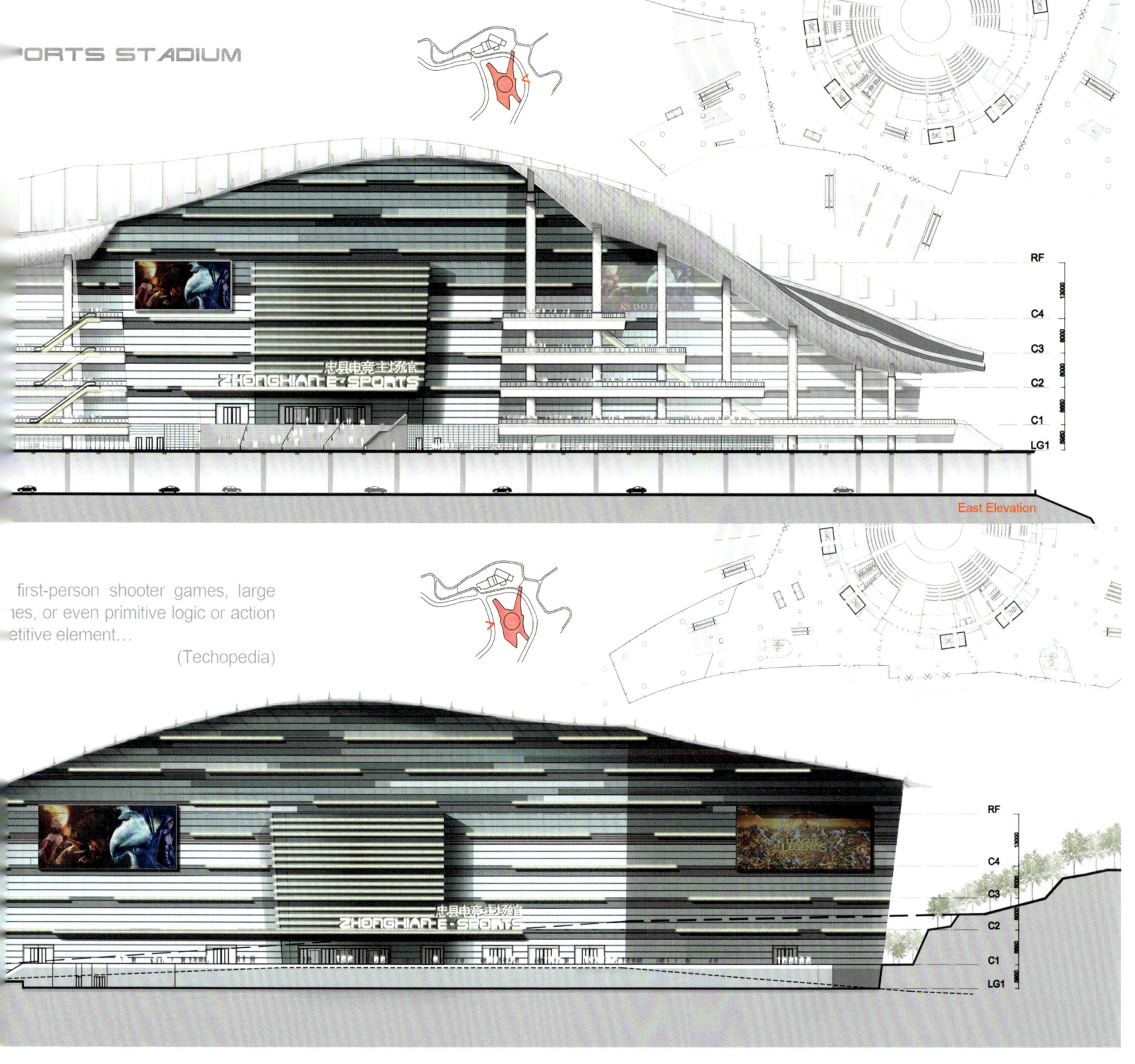

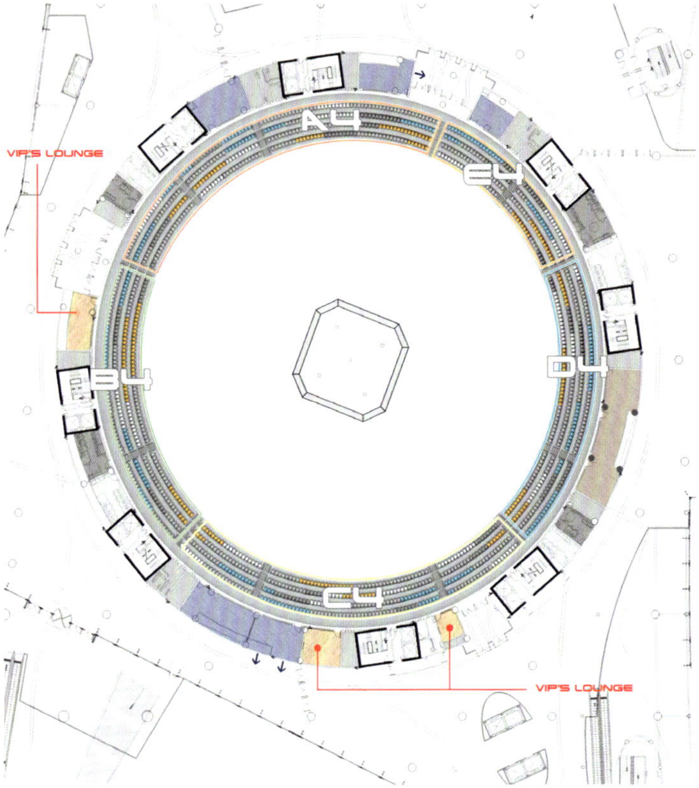

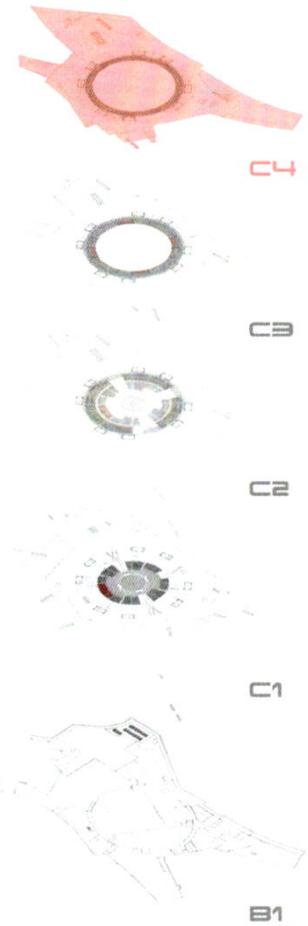

E-SPORTS
[e-spo:ts]

…the activity of playing computer games against other people on the internet, often for money, and often watched by other people using the internet, sometimes at special organized events…

(Cambridge Dictionary)

South Elevation

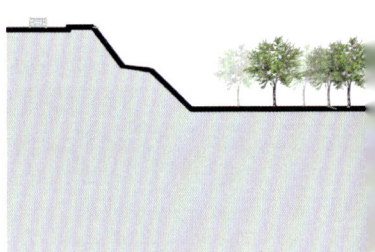

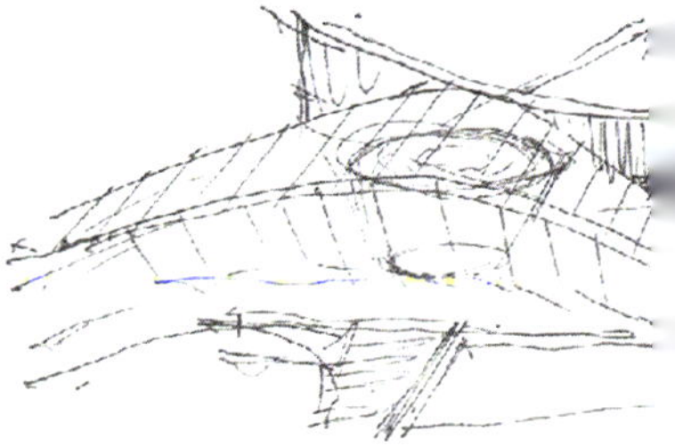

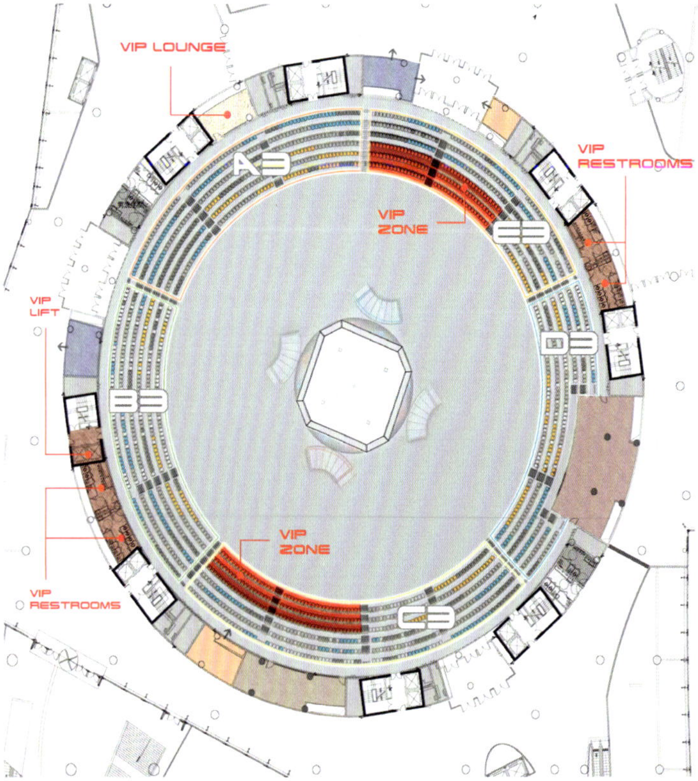

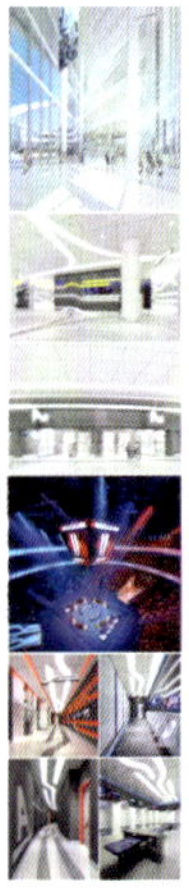

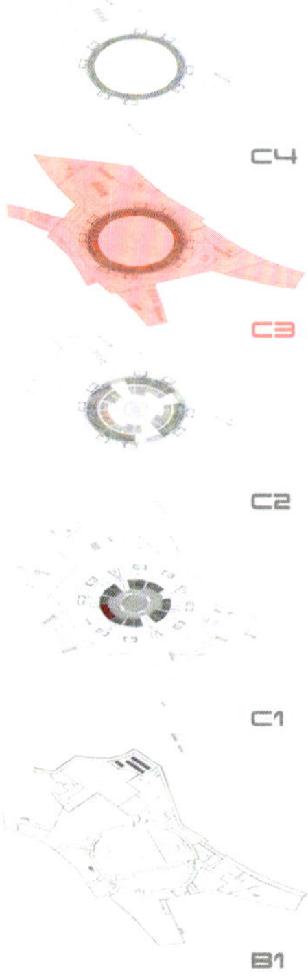

E-SPORTS
[e-spo:ts]

…a multiplayer video game played competitively for spectators, typically by professional gamers…

(Oxford Dictionary)

North Elevation

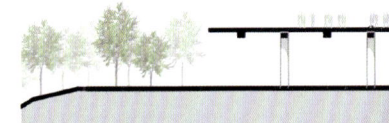

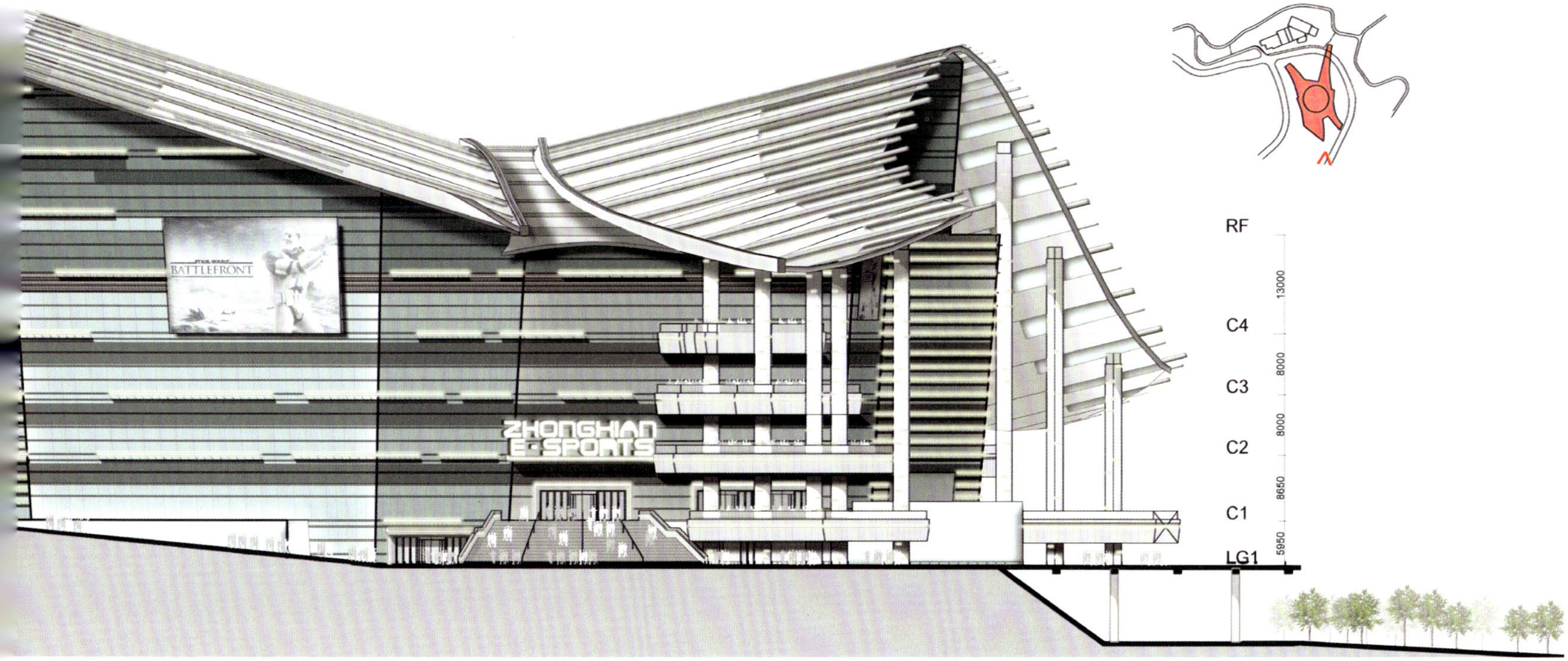
STAR WARS
BATTLEFRONT
ZHONGHIAN
E-SPORTS
RF
C4
C3
C2
C1
LG1
13000
8000
8000
8650
5950

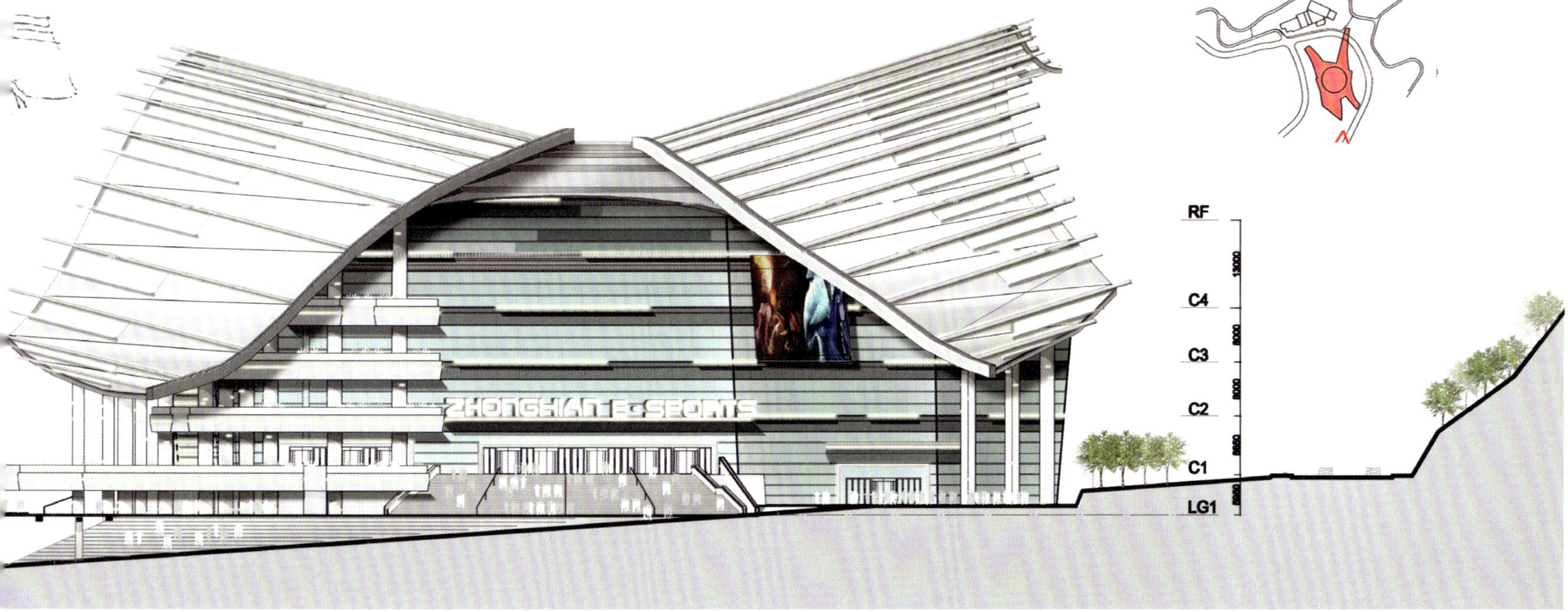
ZHONGHIAN E-SPORTS
RF
C4
C3
C2
C1
LG1

JOCA VAN DER HORST

Onda is a decorative textile piece that purifies the air. It is constructed of fabric with a thin coating based on copper-doped titanium dioxide. When combined with this chemical coating, the textile uses daylight to create a reaction with pollution such as soot and odor, transforming them into, for example, harmless water molecules.

Users can track the air quality by lightly pressing on Onda's fabric. This activates a light that will remain slow and calm while the air is clean, but become quick and frenzied when sensing pollution. Users can also momentarily boost air purification by pressing and holding the piece, lighting the frame to emit ultraviolet light, the most efficient method of rapid air purification.

The work was inspired by research at Eindhoven University of Technology. I stumbled upon a scientific paper about a nano coating that cleans the air silently and invisibly, using visible light. I saw an opportunity to apply this coating on textiles, to create a different kind of air purifier. In the process I considered many form factors and textiles, to find a balance between functionality and user experience.

Onda shows how we can embed technology in a more subtle way in our homes, by turning air purifiers into works of art. With interchangeable textiles and designs, it unites life-enhancing technology with personal expression.

HSC DESIGNS
HILONI SUTARIA - FOUNDER/LEAD ARCHITECT

METAMORPHOSIS OF TECTONIC FURNITURE - THE ROCHEUX CHAIR

The design of the chair is inspired from a micro to macro architectural research to built project. The unfolding facets of the chair are derived from the intensive weather and sun movement pattern calculations for the façade of a residential extension project. The structure of the tessellating planes forming the chair, gives in a self-supporting structure making it strong enough to sit, stand or jump on despite being light in weight. The idea of structural strength comes from the strength of the architectural structure which forms a self-supporting structure. This almost creates an idea of structure that could be replicated through different scales and have the same structural integrity and strength without needing additional supports. The chair is a conversation starter and versatile enough to fit into any space, indoor or outdoor. They can also be kept at multiple angles and multiple pieces of furniture, ie. Seating, Store console etc. We have developed this chair in different materials; Concrete, Wood, Boards, Mdf, Metal and Many more.

DESIGN IDEATION THROUGH DIFFERENT SCALES - S,M,L

Concept of Research: Our design firm's ideas heavily rely on responding to tactile and visual impulses of the user. Testing the concept of ideas to make changes in visual perspectives of people, we wanted to conduct an experiment through built forms translating the design perceptions of Architectural scales into macro and micro scales to see how they influence the end user hedonistically and functionally. For this research we had taken two, Architectural and Interior design scale projects and decided to make products, furniture design and Architectural scale follies and spaces from those. The design ideas have a sense of tectonic fluidity. The idea of this research is to find the design iterations needed to transition into multiple scales, contexts and functions making it the perfect modular solution through scales of Architecture to Jewelry design.

PROJECT -THE SHAPESHIFTING PRODUCT
TECTONIC VERSATILITY
SCALE - SMALL

Adapting a similar idea, the product designed in a micro scale which is adaptable in multiple products. We are currently developing it into multiple products; Magazine stand, Planter, Handbag and shoes.

SYMBIOTIC PARASITE: RESIDENTIAL EXTENSION
SCALE - LARGE

The Architectural scale of the design is an extension for a residential bungalow. The structure is tessellated considering sun path calculations leading to angled windows that bring in maximum diffused light throughout the day without the sun-glare, forming an interesting shadow pattern. The nuance of the new addition preserves the architectural identity of the existing layer and the new layer. This idea stems from the belief of the firm that architecture, like poetry, art or history should be able to make its mark. This philosophy gives rise to the name: "Symbiotic Parasite"

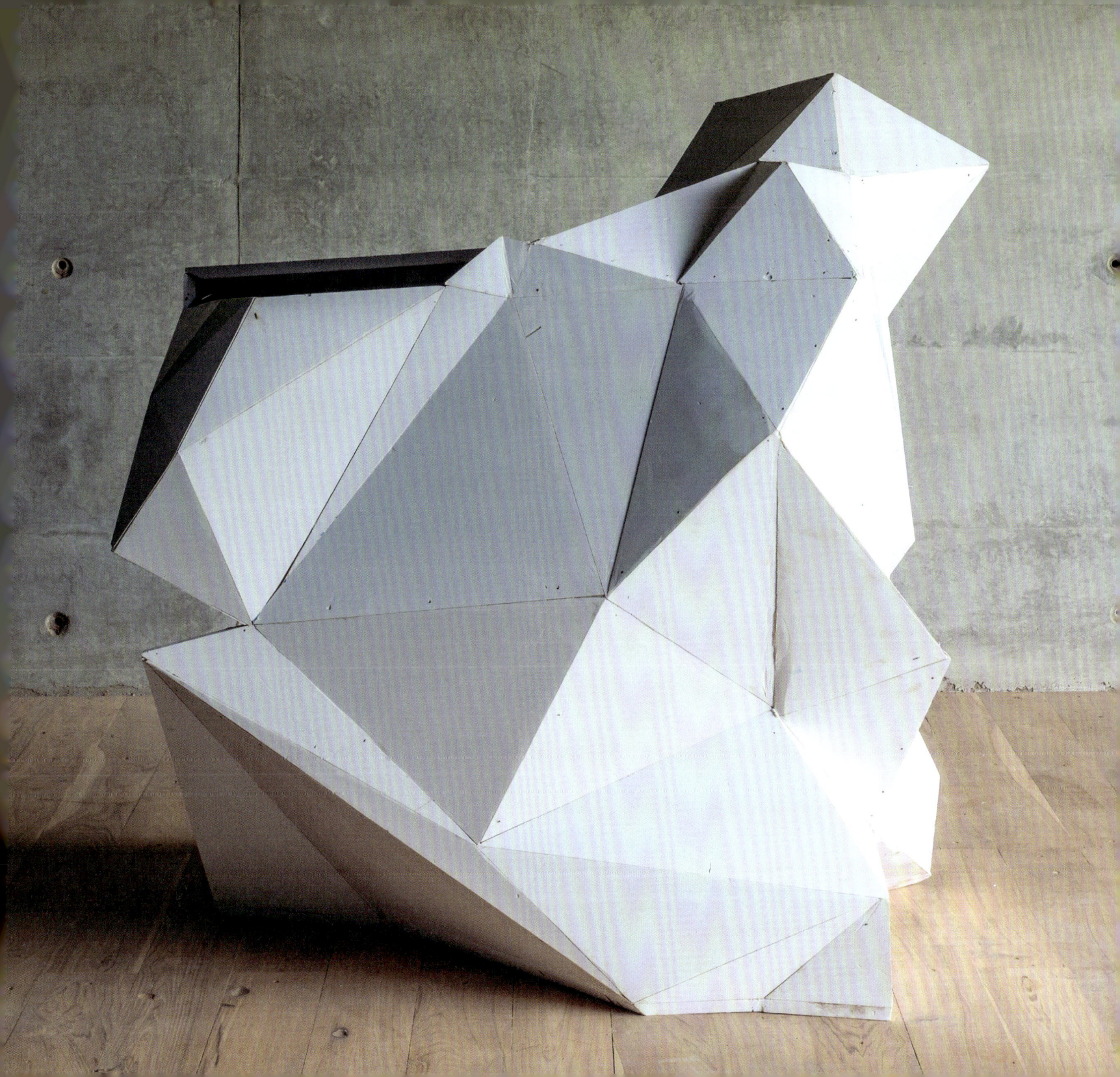

KATALIN HUSZÁR

'Notjustuseless' is an ongoing project aiming to forward a message to ourselves as a consumer society. This design venture plays with the concept of recycling as we know it, bringing it back in a new, refreshing way. It's approach works by changing this dated, old-fashioned concept through interactively inviting the audience into the designing process, opening a door into a modern perspective and consciousness. Consumers have the largest impact on how the project progresses and form the final design with a simple act, by not throwing away their used drinking straws, but putting them in nearby containers designed for this purpose. A QR code - located on the collecting boxes - navigates to the page of the project, allowing you to follow the whole process. The designs' concept shifts and adapts with the change of colours and shapes of the collected straws, resulting in one more level of depth and interactivity with you, the public.

The title of the project aims at showing us its' nuances and double meaning. The words „not just use less" refer to not just using LESS, but at the same time, as you use up new straws, invite you to be reflective and take responsibility into your own hands, when facing the waste from your normal life's material consumption. It does this by stating that used materials are not just USELESS, since every kind can be functional and live again in another form. The design ideally returns to the location, offering consumers a chance to see their not-so-useless waste transform into use again.

MURAAD IBRAHIM

I found out first about Venice Design was when I was visiting Venice and found this gallery I just walked into and found so many beautiful pieces. I remember mentioning to the workers there that I am an artist as well and they took down my information and I few months later I am submitting work. The concept is inspired by the 1980s nightlife scene along with its infusion with art in New York City. In a nod to nightlife legend like Suzanne Barsch, James st James, and Sasha Velour. The words Stay Weird will shine with light highlighting the purpose of the art scene gaging artists and art influencers to stay authentic to their self and creative thinking. The concept along with the piece is minimal and silent. The Alien represents how I view artists and art by, a very intense social construct, artists tend to always outshine and stand out. Just like an alien with a group of people would.

MEGUMI ITO

LIGHTING OBJECT

I was born in Germany, and raised in Japan and already living in Vienna for over 25 years now. Because of that, I think whether my impression of my work is Japanese or Jugendstil, which was very much inspired by Japanese art. Each one is wonderful, and this piece is which I took in and made while imitating both.

When I think of a design, I often imagine shapes that are already set in my mind. Sometimes I just draw my thought on paper and check the image again, or I render it.

In my early career when I started designing, I had the lighting object be made by a carpenter, and I was impressed to get the same image which I had envisioned which was not always the same like that.

In this Lighting Object I see "Clearness and simplicity", like the typical Japanese style and the Jugendstil. I used dark wood because it reminds me of traditional Japanese furniture. Each light turns on individually beneath each of the seven windows. It may be banal to use No.7, but the object is 160cm wide and it is quite long so I wanted to put boxes to make it livelier and give it a kind of typical Japanese "straight and clearness". I feel comfortable with the window proportions as well.

I saw the beautiful location and felt the atmosphere and understood what could make there more and also I wanted to fill the place nicely and sophisticated like the Palazzo and so we chose this object. The object suits the charisma of the space and playing with light and shade to create the right ambience.

Important for me is not only how the object fits into the room, but also what lighting mood the client expects. Because light has a direct effect on humans, similar to music.

*Jugendstil (art nouveau, modern style), style of art at the turn of the 20th century, in Austria closely associated with the Vienna Secession and Wiener Werkstätte. Late phase of historicism and transition to Modernism. The span of Jugendstil ranges from simple household articles to large-sized wall mosaics, from jewellery and glass design to architecture.

Material Wood , Wenge 2004

STUDIO SILVIA KNÜPPEL

We all need them and we all have them.

They help us to keep our homes and offices neat and tidy, the household apliances. All of us own a varity of different cleaning tools, but due to their mostly not very attractive appearances we hide them in dark corners, behind curtains, or inside the cupboards of our homes.

Neutralizer 01 is a cover for a plastic floor cleaning set, a bucket and a mop.

The Neutralizer collection contains different covers for household apliances, like floor cleaning sets, vacuum cleaners, ironing boards, or laundry racks.

The Neutralizers honor the „art of cleaning" and turn these banished but indispensable tools, into contemporary household sculptures, which don`t need to be hidden, but displayed in the living room.

EVA LEVIN

NATURE'S RAW FORM

The piece *Nature's Raw Form* builds on the diversity of biology. Nature's raw form, as close to true-nature as possible. Nature's raw form & shape, a genuine feeling, a authenticity we do not want to influence but let it be as it is. Nature's raw form & shape how nature lives in symbiosis with each other, for "maximum potential" of experience. Man is originally one of Nature's raw form & shape, we must not forget the senses we have received, in order to experience Nature, in its true form. Nature's raw form & shape, means that Nature via evolution, given us the solutions and variants that exist today - To link to that evolution, to remember where we come from originally: Nature's raw form & shape, it is organic.

Man's minds are burst out of nature - we come from there and for thousands of years our minds have become more blurred. To go "back to the origin" with nature as the source of inspiration, creativity, structure, experiences and well-being. Nature has a built-in balance and harmony. Back to the origins that we humans once were, where our minds are sharpened and used.

LUMNEO
BENJAMIN MÉRY

Benjamin Méry concretizes his dream of giving back a contemporary image to the neon. Attracted by his vintage style, he wants to integrate it into a mirror and create a « trompe-l'œil » with a depth effect like artists of the years 60/70 who worked the neon but modernizing the object, giving it a true mirror function and integrating all the technicality in a frame with absolute clean lines and refined style. Benjamin likes to explore all capabilities of 3D modelling, 3D printing to design the mechanism used in its Mirror Lamp. However, he associates this modernity with traditional know-how such as those of glass-blower, in aspiration towards new possibilities.

He imagines a bright neon mirror and launches LUMNEO, a concept of lamp-mirror art design, multifunction, declined in various colours and creator of atmosphere. "Warmer than a LED, the neon light offers a presence. Once switched on, the bright perspectives are reflecting to infinity and transform the mirror into an artistic object, almost alive in the room, "explains Benjamin Méry. In addition, the luminous intensity of the neon modular to the envy, allows to create the desired atmosphere. Strong of a variation of shades and of a trompe-l'oeil effect, the object becomes a vibrating artistic piece, pushing the boundary between art and design.

Produced on order, each mirror lamp is customizable to accommodate specific requirements (wood species, dimmer button, neon colour...) and to meet growing aspiration for unicity.

For Venice Design 2019, Benjamin is happy to present its TRYPTIC MINI in an installation of three MINI: Turquoise, Coral Rose and standard Blue in Mahogany wood and tempered glass. The combination of three MINI in installation reveals all its artistic rendering with subtle marriage of colours. The dimmer buttons on the top provides a playful effect allowing the public to appropriate the object.

In contrast to the version MINI which can be positioned vertically or horizontally, the other declination of the Mirror Lamp called MAXI (large model 180 cm high), goes so far as to make disappear the dimmer and switch on the back of the mirror for a perfect mirror illusion in the off position. When switched on the MAXI mirror convenes the mystery and creates the illusion of another space. The "trompe l'oeil" plays full and challenges the human eye. The link between light, colour, space and their perception are at the centre of his work.

Benjamin Méry decides not spare on the quality of this product hand made in Italy with neon tubes made by a blower glass, wood worked by craftsmen and high-quality glasses. His philosophy is to resist globalization and uniformization of design by favouring local manufacturing and traditional work.

Thanks to its infinities of combinations, Lumneo offers uniqueness of scenography to satisfy all dreams.

Our thanks go to Atelier GOHARD for supporting financially Lumneo's participation to Venice Design 2019.

LYK CARPET X UNIQUE FACTORY

REINTERPRETING THE CRAFT OF HAND-KNOTTED RUG MAKING – LYK CARPET X UNIQUE FACTORY

With Pouf-Characters began a collaboration between LYK Carpet and UNIQUE FACTORY, a Berlin-based manufactory for living design. They both quickly discovered this was going to be a great partnership for creating the objects. Working closely together, giving adequate time to preparation, enabled them to create pieces to the highest standard of workmanship. The manufacturing team of UNIQUE FACTORY brought all their experience of materials and scale to the development and construction of the objects, and in addition, the very precise focus necessary for hand-sewing these extraordinary carpets by LYK Carpet. The whole collaboration was inspiring and delightful and is surely going to be continued into new projects.

UNIQUE FACTORY Berlin is a manufactory for living design, founded in 2010 by Mensud Bjelosevic and Mirza Music-Zander. Their core business includes upholstery, furniture tailoring, and made-to-measure home textiles. They are dedicated to combining tradition with modernity: " Individuality is what we aspire to. Faceless mass-produced items from nameless mega-ranges leave little room for objects with a unique character. We want to offer alternatives, with sustainable, worthwhile products that stand out due to their unique-ness and charming individual identity. Workmanship, experience, and intuition are the foundations of our manufactory, we are driven by our passion for perfection and the thrill of trying new things."

For her label Lyk Carpet, Mareike Lienau pursues her mission to reinterpret the craft of rug making, and to kindle people's appreciation of it by designing one-of-a-kind rugs, objects, and textile wall-pieces, which are then crafted by hand in Nepal.

Mareike Lienau is the owner & founder of Lyk Carpet. Her approach asks, 'What is contemporary about craft?'

"Many people don't really relate to products anymore. Mass production has led to today's throwaway society. I want to contribute to raising people's awareness, prompting them to think about products: how they are made and the resources involved; to generate values. I decided to do this through supporting the preservation of a cultural heritage, and by designing authentic objects. Contemplating and experiencing this artistic craft leads to new insights and points of view and enriches communication. With my textile posters and Pouf_Characters I am inviting people to acquaint themselves with a value creation chain, and to think about global responsibility."

TAHIR MAHMOOD

Antidote art & design returns to Venice Design for the second time, showing two design pieces by Toronto based Pakistani designer Tahir Mahmood. The two pieces on display at Venice Design are *Rustam Dip Pen and Nib Holder Set* and the *Killi and Taal clothes hangers*.

The Rustam Dip Pen and Nib Holder Set is made from natural Rosewood and named after RUSTAM-M-ZAMAN, a well known Wrestler from the Indian Sub-Continent, this Dip-Pen set is a tribute to the 'Pahalwan' (wrestlers) of by-gone days, who were as much a part of the local culture as the historic buildings they wrestled amidst. Dip pens date back to the early 19th century and it was the first step toward turning the pen from a handmade tool into a manufactured commodity. A unique tool for any writer, artist, cartoonist or anyone who simply appreciates writing beautifully in today's world. This interactive installation will invite the visitors to use the pen set themselves for writing notes or letters to loved ones.

Made from natural Rosewood, the Killi and Taal clothes hangers by are functional yet truly an aesthetic pleasure to have in any surrounding. The Taal hangers are inspired by the architecture of the Moghul Era with their cylindrical minarets and the ribbed ends. The tips of the Taal have been embedded with a semi-precious gem 'Aqeeq' or Carnelian stone, believed to be symbolic in protecting individuals. The classically round shape of the Killi with its pastel colours is a contrast the Taal. The smooth rounded edge provides ease of hanging while its colourful face appears decorative even on its own. Using the ancient technique of lathe turning to apply resin, Mahmood works with the artisans in Pakistan who have perfected this art and have used it for generations.

Radically simplified forms and angular silhouettes are typical of Mahmood's work. From his early days of designing he showed an understanding of how mass-produced items could be reconciled with aesthetic details and purity of materials. Mahmood's awareness of the German design school Bauhaus was not far removed from his own desire of finding a visual language relevant to the social and cultural environment of his own upbringing.

Antidote art & design represents Tahir Mahmood internationally and these design pieces are available to buy online.

Based in Dubai, Antidote is an art and design platform representing talented artists and designers, both emerging and mid career, from around the world with a special focus on the Middle East North Africa South Asia region. Antidote supports the careers of its artists and designers by guiding them to appropriate residencies, special programmes and biennale placement.

Antidote was the organizer for the first ever Pavilion of Pakistan at the Venice Architecture Biennale in 2018.

M.I.C.A.T.

MICAT (Moroccan Initiative for Craft, Art and Technology) was established by IFASSEN and Aziza Chaouni Projects as a social enterprise aimed at creating a plastic recycling and design workspace.

IFASSEN, a brand founded by Faiza Hajji in 2006, provides revenue opportunities to rural Moroccan women and protects the environment. It began with weaving worn plastic bags into traditional baskets and has since developed, featuring creative design, innovative materials, and artisanship. IFASSEN has a wide network of women's cooperatives with which their team works, and is committed to sustainability.

Aziza Chaouni Projects (ACP) is a design firm with more than 10 years of experience in Morocco, and is dedicated to promoting sustainability in construction and collaborative design. ACP has developed projects all over Morocco for disadvantaged communities ranging in scale from urban furniture to urban projects such as the daylighting of the Fez River.

IFASSEN and ACP, led by two Moroccan women and close friends, believe strongly in collaborating and sharing their expertise in design and social entrepreneurship in order to achieve two objectives: diminishing plastic waste by giving it a second life and helping women's cooperatives reinvent their trade by proposing new projects with added value. After a year of brainstorming Aziza and Faiza have developed MICAT, a recycling workspace to tackle the problem of plastic waste while enhancing craft production for women's cooperatives in Morocco.

As a plastic recycling and design workspace, MICAT will collect plastic waste, shred it, and transform it into filament for 3D printing, in turn allowing for the production of a wide variety of well-designed craft products. For the time being, MICAT is focusing on lamps, mixing traditional Moroccan weaving and 3D printing. This production will have an immediate benefit for the women's cooperatives, with an enhanced revenue stream. This is because the 3D printed frame decreases the amount of time spent weaving and the labor necessary to construct a lamp. Additionally, the product itself will be sold on a different market than the traditional crafts that were produced by cooperatives and then sold locally.

MICAT will create new work opportunities for locals by hiring them to gather recycled plastic for the shredder. MICAT will also offer training opportunities for youth by bringing a specialist in to teach the team how to run the equipment. In the future, as new products are developed, more income opportunities will open up for the partner women's cooperatives. The collaboration between ACP's designers and the women's cooperatives will allow an exchange of skills in both directions. Rural women will feel empowered because they will be exposed to new technology and will be given free rein to let their creativity expand. Awareness of plastic recycling will also increase through the experience of seeing how plastic waste can be transformed into a unique craft product.

TERESA MOORHOUSE

Teresa Moorhouse is a graphic designer and illustrator from Finland.

I have a background in fashion design and graphic design. I studied in Paris and in Finland. I have designed successful designs mostly in Finland, Japan and South Korea.

I am very much inspired by the Finnish nature, Mythology embedded in Finnish culture, however linked in today's world and its needs for joy, happiness, imagination and esthetics.

At VENICE DESIGN 2019 my work reflects Nordic imaginary textile prints with nature themes. The big motif, Nanuk (designed for Marimekko), the polar bear whose background has suddenly become a flower field. The ice has disappeared and the bear is surprised, wondering what has happened to his ice and snow. He is startled and gazing deep into our eyes.

Kaunis kauris (beautiful deer in Finnish, designed for Marimekko) has stopped to stare at you in the mystical Finnish forest. This is the moment I have always hoped to happen while walking in the forest. The magical, very short moment of silence and the gaze of the beautiful deer. This deer represents the magic of the forest. The forest is a sacred place for many Finns. Plants and nature have become more and more surreal and nature is changing in ways nobody knows. The meeting of living things in nature, wild animals in particular, is like a holy moment for me, and this is what I want to express in this design.

I like to design with a sense of playfulness, lightness, and a graphic touch. I feel people need joy and happiness in their lives. Which I try to bring with my illustrations.

The colours are often vivid and strong. In Finland most of the year days are extremely dark but in summer light is overpowering. We even have nightless nights, with light the whole day and night. I think this also reflects in my use of colours. This atmosphere encourages contrasting colours with greyish hues.

Myllymäki, the abstract pattern with simplified blueberries from the Finnish nature. This is the food of the Finnish wild animals and very good for people too. I like minimalistic style combined with decorative patterns.

The bear rug, "Karhu" (designed for Mum´s, Finland) is hand made in India and designed in Finland. The bear is a symbolic animal in Finland and embodies many mythological stories from the folk history. The "Karhu" is designed in a naivistic style. Karhu is an empathetic and lovably design for homes everywhere.

The material of "Karhu" is of very ecological silk bamboo which adds softness and luxury effect to this warm and strong animal of the woods. Karhu is my newest design in Venice 2019.

Together all of my patterns create a dialogue with The Venice design 2019. We all should take care of natures treasures!

Special thanks to Suomen Kulttuurirahasto for making this exhibition become true.

BRIAN NAEYAERT

'THIS AIN'T NEW YORK, THIS THE BRONX'

Hey, Mom, these fools busted us for dancing, can you believe that?

Dreamers will take risks

When are you going to stop writing on the walls and make some money, when are you going to stop tagging the subway cars, when are you going to make your son legitimate ... legitimate shit

I'm a dreamer
              B.

Hiphop/breakdance in 1980s South Bronx, New York, is central to this collection. To elaborate on the theme, I used image, colour, and inspiration from "Beatstreet", a 1984 feature film that truly put breakdance on the international map. The explosion of movement that is characteristic for breakdance and the contours of the dancing formed the central axis of building up the silhouettes. For the materials, I chose a range of fabrics going from fine silks to heavy synthetics, a mix of materials to highlight the constant movement and depth in dancing. As in dance choreography, the detail in the garment and finishing refers to technique in dancing, combining movement with a sense of great detail and form. Colour-wise, I balance everything out by using strong black, white, gold and multi-coloured silhouettes in contrasting ways. Going back to hip-hop and breakdance in the 1980s, I combined the typical sportswear DNA with different high fashion approaches.

This installation is part of my graduate collection 2017 'This ain't New York, this the Bronx' at the Fashion Department of the School of Arts/Royal Academy of Fine Arts (KASK) in Ghent, Belgium. This collection was a creative expression of a story I told within the guidelines of a master assignment. With the generous support and guidance provided by Liesbeth Louwyck; Eva Bos; Marina Yee; Hugo DeBlock; Bram Jespers; Ann Lannoo & Adel Vanlerberghe.

NIVASARKAR CONSULTANTS

I believe that for any product, especially furniture, physical and emotional comfort for the user, is core to the design being relevant. Physical comfort can be achieved by considering usability, appropriation of the right technical specifications and right technology. To achieve a sense of emotional comfort is more challenging. One needs to study social science, cultural anthropology and tradition, which enables the designer to create an aesthetic that fits seamlessly into the environment, the product was designed for. This process breathes personality and life into everyday objects.

My explorations in furniture design have focussed on experimenting with natural materials, wood (mainly recycled wood), natural rope (made from the Hibiscus Cannabinus plant) and natural cane, with a preference for organic lacquer to finish the piece. My effort is to be as sustainable as possible, while also restricting the use of machines to a minimum. Working with hand tools, exploring traditional skills and crafts with a contemporary flavour, are key to my philosophy.

In the realm of furniture design, designing a great chair is the ultimate challenge and a gratifying experience. Designers across the world have designed many acclaimed chairs, however, even today, every furniture designer aspires to design at least one iconic chair in their lifetime.

In my career spanning four decades, I have explored numerous options of chairs that cater to different needs. Of these, some provided me with great learning experiences of what not to do, while others merited further exploration and refinement, because the journey towards excellence is ever evolving.

The *'Presitge'* works on a volumetric form, which lends it a powerful presence. The frame, seat and backrest are all made using one section of wood, which also renders the chair convenient for mass production. The seat and backrest are created by joining sections together and scooping material in a manner that not only makes the chair comfortable, but also enhances the natural beauty of the wood. Exposed joinery highlights the simplicity of the design, complemented by subtle ornamentation of rosewood inlay that adds an understated elegance.

'Spine' has a more or less similar approach, but is an exploration of a more playful form. Bolder ornamentation through inlay, adds to the personality of the chair.

The *'Ecological Chair'*, is so named because of the materials used in its construction. The frame of the chair and foot stool are made out of recycled wood. The integrated seat and back are woven with *bhindi* rope (Hibiscus Cannabinus), which is a rope made from a locally grown plant fibre. *Bhindi* rope is a great material for tropical climatic conditions. Being hand woven, there are a variety of designs and patterns that can be explored for the weave, changing the personality of the chair without altering its design and construction.

JINHWA OH

Jinhwa Oh is a graphic designer currently based in New York. She was born and raised in South Korea, and moved to the U.S. in 2016 as a Fulbright scholar. Her work has been recognized and published internationally by the Type Directors Club, Red Dot Award, and Wallpaper magazine. Currently she works at Base Design.

She has been working as an in-house and freelance designer for more than four years creating a variety of things: visual identities, printed matter, websites, and environmental graphics. Previously she worked at Pentagram, Imprint Projects, Samsung Electronics, and CJ Foodville.

PROJECT TITLE: 360° ALPHABET
DATE: MAR. 2018

360° Alphabet challenges conventional dimensionality in typeface. Every letterform has a three-dimensional figure which shows different visuals depending on its angle. The inner lines inside of each letter rotate more on the z-axis as they are closer to the center of a letter. Due to the regularity in transformation, all the letterforms look very similar when they turn 90 degrees. As a set, the alphabet has both regularity and irregularity in its form, residing in a three-dimensional world.

PEMARA DESIGN
PER PLOUG

Pemara Design is an Irish based furniture design company with roots deep in the Nordic furniture traditions, focusing on the development of own unique designs. The company was founded by Danish designer Per Ploug, who has worked in the furniture industry for over 20 years as a kitchen designer. Combining his many years of experience, his engineering skills and his knowledge of good interior design, made him form and create a unique dining table. Other designs, such as bench, chairs, barstool and side table since followed and are all part of Pemara Design's collection of well-crafted, beautiful and inspiring design.

Pemara Design was chosen to show three pieces in Venice Design 2019, the three-legged Cauda Chair, the Veizla Side Table and the concept Veizla Mirror commissioned from Irish mosaic artist Laura O'Hagan.

CAUDA CHAIR

An iconic dining chair is one of the ultimate challenges for a furniture designer, since it has to fulfil an array of important design criteria... It must be strong enough to withstand years of use, light to be easily moved around. It must be comfortable to sit on for long periods of time and support you ergonomically well, and last but not least it must be aesthetically pleasing to look at from all possible angles. The Cauda is the second chair in Pemara Design's collection, a further development from the Graphium Chair.

VEIZLA SIDE TABLE

Pemara Design's collection of furniture began with the Veizla Dining Table. Veizla is an Old-Norse word for a feast or a banquet. It was designed from the philosophy that dining should be a highly social and collective experience, always inviting and a table should always be inspiring both in and out of use. This philosophy is evident in all the pieces in the collection. The Veizla Side Table is elegant yet playful, and most definitely inspiring. The form of its table top makes it ideal for placing either freely or up against a wall while giving room for a set of chairs to angle towards one-another for a heightened social experience. As for all Pemara Design's pieces, made from wood sourced only from sustainable forestry and finished in natural environmentally friendly treatments.

VEIZLA MOSAIC MIRROR

For the 2019 Biennale Pemara Design has fostered the concept of creating a unique and individual series of mirrors in the image of the round Veizla Table with the three "beaks", they commissioned mosaic artist Laura O'Hagan especially to make for them the first one-off art piece. The piece "Late Summer" is inspired by the colours of deciduous Nordic forests, when the leaves on the trees are beginning to turn and the first few have just started to fall off their branches. It is the intention to commission more such pieces in the future, with various themes and from artists working in different media, in the same overall size and shape. Pemara Design envisage creating an iconic and most individual line of mirrors that truly connects art and design in a simple yet powerful way.

PY MANUFACTURE
YANN PÉRON

The object is born with the merger of intelligence related to the shape we will give it, the material's beauty and the meticulous work of a conscious gesture. Not only mind and body unite together to give a soul to the object, but also to spread an inalienable and immeasurable value that influence daily life. It becomes alive and comforting. Even at rest, it contributes to your interior world, to the spiritual development of the occupants of the house.

Emblematic and symbolic object, this eggcup materializes the chicken-and-egg metaphor into copper, sublimated thanks to the accuracy of the brassworker work. This geometric and fractal shape is proportionate to the « Flower of life » that was born from a circle divided into six equal parts, that themselves create six other circles, etcetera, etcetera. As well, this eggcup shape allows us to use it in both senses, signification that eggs come from chicken and vice versa.

Detachable from one another, theses two parts of « eggs-storage » showcases eggs at hands on kitchen shelf. The « male » removable part is used as an eggcup of presentation. Golden brass and marble are in connection like sun and moon to create only one harmonious object.

QSTUDIO

LUM

Whenever you need to highlight an important wall, be it a building, an office or a house, you first start seeking for a coating that gives life to that space, then you investigate on how to illuminate it and then you hunt for external products that would illuminate it from above, below, the sides or even from behind.

That's why LUM is born, a copper cladding that is made up of modules with light, which combined with the basic modules generate different compositions, according to the aesthetic and lighting requirements of each space, achieving an exclusive coating that highlights your space and that also takes charge of its own lighting, with no need of external light.

The combination of the different modules, plus the different colors of the oxidation of copper, make it an extremely versatile and exclusive product, since there is no wall equal to another.

ENVIRONMENTAL

We use sheets of electrolytic copper LEC, which does not require the process of copper rolling. That means a significant saving of 80% of carbon footprint. In addition, copper is a 100% recyclable material.

SOCIAL

It is manufactured in our workshop at the Colina 1 penitentiary center, where we train and give work to people deprived of liberty, allowing them to reintegrate into society and work.

ECONOMIC

Lum gives added value to our main raw material, as it helps not selling raw material but a finished product with added value.

REVOLOGY

We have limited resources on this tiny, beautiful planet we call home. As creators, our choice of materials and how we design will dictate how fast we run out of those resources – or not. At Revology we believe in the "or not". We think in circles not lines. Our design process starts with choosing the right materials – renewable, recyclable and non-toxic and let them define the choice object based on their unique characteristics. Next we design objects to last, making them capable of evolving over time and ensuring that they are easily repairable. Finally, we find ways to recover materials at the end of their natural life and put them back into the cycle. A bit like nature does already with materials it uses. A circular economy starts with circular design.

That's why the chair, design #1, is made from plants. The legs and back are made from flax fibres, which are woven and intermingled with a resin to form a composite structure. Sourced in Europe, flax takes zero water other than rain to grow, stocks carbon, takes 90 days from seed to harvest and is infinitely renewable. It's also non-toxic for workers and smells like gingerbread rather than burnt oil when being moulded. As a composite, it has the same strength as fiberglass and is lighter than carbon fibre and we can recycle the material into other objects. We've teamed it up with a seat made from sugar (don't worry – it's not the sugar you put in your coffee!). The bio-resin is a highly technical product made from plant-based isosorbide that is impact and scratch resistant – made to last. We've designed it to evolve. The seat plate and lumbar support come off with a simple twist of an Allen key. That's where we need you to make the cycle work. You send the damaged part back to us, we give you a discount on the new part that covers the cost of postage and we send you a new seat. We can then put the old one back into the system.

We invite you to take a seat, pause for a moment, think in circles and ask yourself "where do I sit in nature?"

Next challenge? Design the lightest, most sustainable urban c-bike ever. Why? Light, strong, impact resistant. Just makes sense. Follow us to find out more.

LOU VAN 'T RIET

After realizing that people don't really look at art anymore but wander from one work to the next, instead of engaging with the art, Lou has been working on creating new ways of interacting and experiencing art. She has taken art out of the context of traditional galleries and created "triptych_1".

Triptych_1 is an artwork where one can no longer be passive.

Triptych_1 is a graceful interactive artwork that gives people the possibility of seeing something different each time they interact with it.

It consists of four different artworks in one. Every time people open or close one or both sides of the artwork, it becomes something totally different - due to the shapes and colors that appear or are hidden.

The artwork consists of three geometrical enameled parts, elegantly architectural, which rotate onto the sides of the central part. It has been produced in collaboration with Emaillerie Belge.

Lou van 't Riet wants to change our relationship with works of art by using art, design and/or architecture to enhance the experience and environments in which we view them. Triptych's concept is a sensory approach that allows us to experience art on a completely new level, because she challenges the unwritten rules that you can't touch, get close to or take an active role in your relationship with the artwork.

Her work is an attempt to make art accessible, enjoyable and understandable for both people who don't usually appreciate it and those who do.

Lou van 't Riet is a Belgian multi-disciplinary artist and has been passionate about art, design and architecture since birth. After having lived for a few years in New York, she has been back in Brussels since September 2018, where she has opened "GANG atelier". A new creative co-working space where she works on various artistic projects at the same time, while being surrounded by other creative people.

SELEK

Balance Mirror acknowledges all the elements that compose a conventional mirror typology and mutes them down into a pure form. The product is designed to be a sculptural object which could serve as a reflective surface when needed. It aims to provide a silent environment by communicating its primary function as little as possible.

Objects constantly communicate with us. Sometimes this communication between human and products becomes overwhelming. A mirror tells us that it is a mirror through its frame, feet and other features. With Balance Mirror we aimed to silence this communication as much as possible and stripped the mirror from its secondary functions such as frame and feet. This resulted in an object that carries minimal amount of mirror semantics. One could see it as a mirror only when they need it, else it is a pure material that is sculpted.

The trimmed cylindrical geometry facilitates the center of gravity to bring the product to a balance. It also helps forming an angled and a perpendicular reflective surface on the two sides of the product. The free standing object is made of pure aluminum.

GISELA SIMAS

A contemporary seat that brings together the legacy of Brazilian modernism, one of the richest periods in art, architecture and design in the legendary South American country. Gisela Simas introduces the exclusive chair and ottoman Bia at VENICE DESIGN exhibition.

Made of oak wood, using the expertise of the Portuguese company Época exclusively for the Gisela Simas brand, the furniture is characterised by its vigorous design, singular and straight lines, 45-degree angled feet in wood and metal. It displays a "personality" that recalls the famous artistic movement that exploded in the first half of the 20th century. The chair and ottoman are covered with waterproof burel felt, providing a sober and sophisticated nuance.

Bia - a tribute to the nickname of the designer's sister Beatriz- has a poetic meaning. "Its structure is inspired by the shape of a boomerang, an object that recalls infinity, the freedom. Freedom that has enabled modernist artists deploy to disrupt tradition. As the boomerang returns to the starting point, Bia is a piece of furniture that, despite its modern and international style, and brings back memories from a significant period for Brazilian design. This was not intentional, it was only after the creation of Bia I perceived how much this influence had permeated into the spirit of the project", explains the designer.

Gisela Simas believes that one of the aims of design is to interact. "For me, wood is a passion. I conceive furniture with the aim of making it "talk" with the users, offering a special feeling to the touch, having a leading role in any space where is. When I designed Bia, the intention was to create a visually and structurally lighter seat, easier to move, compact and comfortable, an invitation to sit in it and 'live' it", specifies Gisela.

Gisela Simas boasts more than 20 years of experience and is known for combining innovative design with dynamic production techniques. Her work reveals suggestive shapes, elegance, and fluid movement. The Brazilian designer runs studios in Rio de Janeiro and London, where she lived for ten years and earned masters degrees in furniture and industrial design at Central Saint Martins.

The talent of the Brazilian designer has captured the attention of renowned galleries and exclusive shops, and it has been exhibited at international events.

MIKAELA STEBY STENFALK

Collective Collection: Sistine Chapel is a physical re-construction of the original chapel, using crowd-sourced images from social media and photogrammetry.

In a time when the experience of the physical world often happens first through images circulating on the Internet, the digital image has significantly gained value. As Hito Steyerl stated in Politics of Post-representation (2014): "Social media makes the shift from representation to participation very clear: people participate in the launch and life span of images, and indeed their life span, spread and potential is defined by participation." Within this condition the 'crowd' determines the force of the image, not the professional.

Social media platforms such as Instagram can now be seen an open-source, digital archive of public space; collectively gathered, shared and remembered by the public itself. The visitors have transformed from passive spectators to virtual storytellers, as they upload the narratives of their experiences onto social media.

The Sistine Chapel is part of the Apostolic Palace, the official residence of the pope, in Vatican City. The chapel is home to many pieces of art including statues, tapestries, and paintings by Michelangelo Buonarroti. One of the most famous attractions is Michelangelo's ceiling paintings including the Creation of Adam, as well as The Last Judgment behind the altar.

Despite the strict prohibition of photography inside the Sistine Chapel, the number of images on Instagram are increasing rapidly. Digital categorising systems such as hashtags—a word or phrase used on social media to identify images on a specific topic—these images can easily be gathered. At the time of the production of Collective Collection: Sistine Chapel, the hashtag #sistinechapel had accumulated 118.615 images. Each day, approximately 30 new images are added to the hashtag.

What is the collective memory of the Sistine Chapel? Collective Collection crowd-sources one day worth of images (i.e. 30 images) from Instagram and re-constructs the chapel through photogrammetry. Mimicking a three-dimensional scanner, the public collectively present a fragmented, digital paraphrase of the original space. Simply explained, photogrammetry works better the more images are available: Areas of the chapel which were photographed frequently become clear and detailed, meanwhile areas such as the floor is completely missing.

The method works in a forensic way, which allows us to trace back a moment in time and space, through the lens of social media.

In the model, one can clearly read the gap between a physical reality and a digital one. And it is not gravity that poses the largest challenge for the digital architecture, but to be forgotten, un-liked and unshared. When images of our physical reality are broadcasted, circulated and exhibited as a whole online, even though they only depict selected parts – what is left behind? And is it important, or are we collectively deciding what's worth remembering?

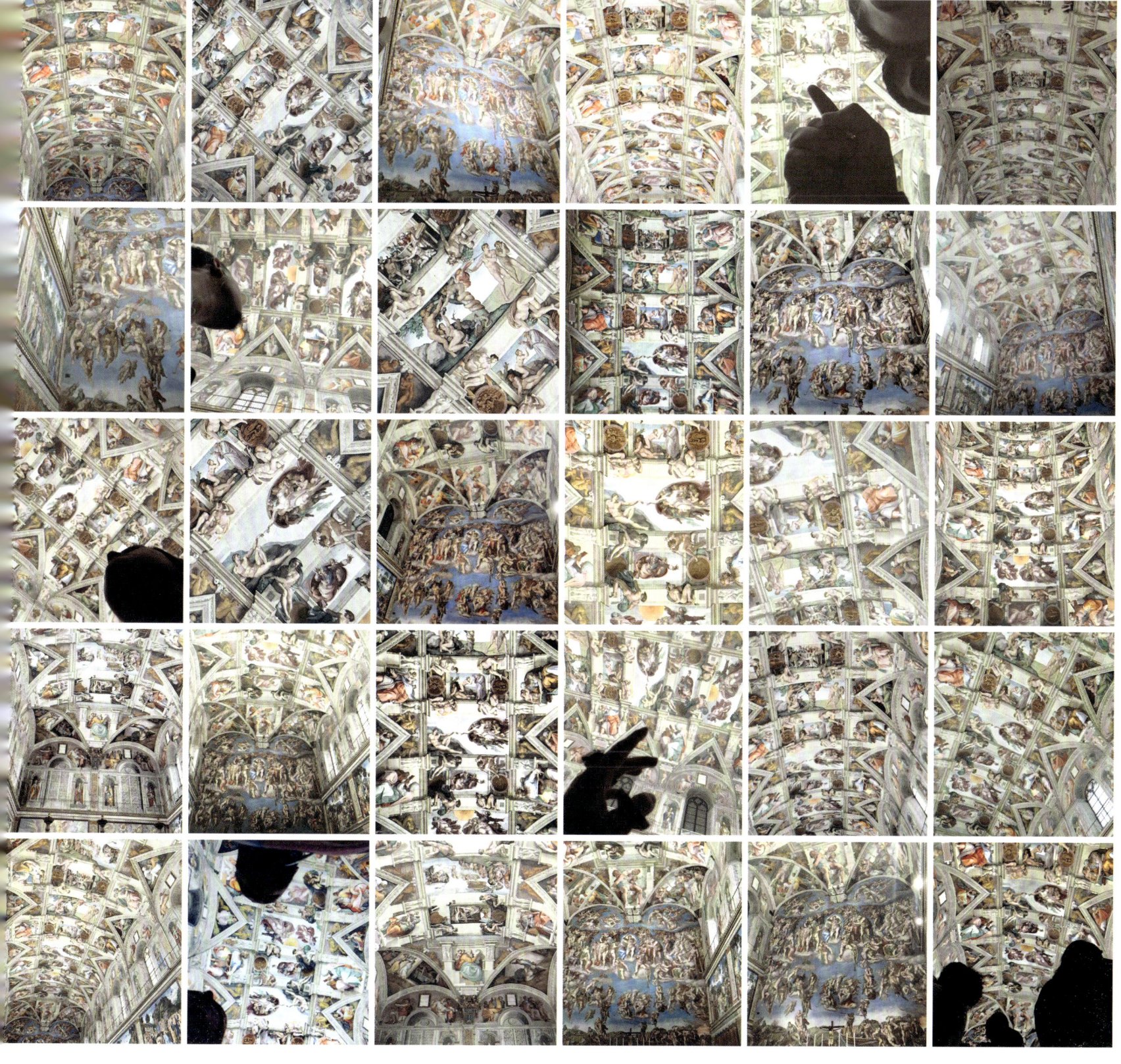

J.M. SZYMANSKI & ALEXANDRA KOHL

LAMP IN CRAVED IRON AND CAST GLASS

Hanging Fixture No. 1 is a monumental ceiling light fixture. With a length of 3 meters, the fixture is suspended by a series of hand carved steel links. The links meet a central cast iron sphere that produces an enormous visual weight. This perfect sphere holds two cast glass and steel arms that are asymmetrically positioned. The intense and dramatic horizontal arms juxtapose against this perfect sphere. Light is cast upward and downward through the arms and out the heavily textured cast glass. This texture within the glass catches the light and pulls it through to the outside.

J.M. Szymanski combines dramatic geometries with tradition materials, steel and glass. The result is a modern chandelier.

J.M. Szymanski is a maker & designer of one of a kind furniture and objects. The studio introduces radical new shapes and explores new materials and textures. Each work demonstrates a fascination with severe geometries, raw materials, and unique forms.

J.M. lives in New York City where he designs and fabricates his work. His design education began in his early childhood where he and his family spent several years in Nepal. Later, in his early 20's, he moved to Spain where he studied the designs of Gaudi, visited the Alhambra, and made frequent trips to Morocco. JM returned to the USA to study interior design. He studied at The Fashion Institute of Technology and at The School of Visual Arts in New York City. He went on to design for William Sofield.

JM has more than 15 years of experience in interior design and has a strong knowledge of the decorative arts.

BENCH IN CRAVED IRON AND HORSE HAIR TEXTILE

Alexandra Kohl and J.M. Szymanski are honored to present a unique design inspired by the rich history of the decorative art in Italy. Horse-hair textile, blackened iron, and Italian marble are combined to form a dynamic juxtaposition of elements. The design combines the traditional craft of a weaver, a blacksmith, and a stone sculpture into a modern context. The artists aim to pay tribute to the history of these unique crafts. Horse-hair, a medium used for centuries, is handwoven on a floor loom by Alexandra Kohl. Contemporary geometries are woven into a pattern that results in a contemporary tapestry which is used as the upholstery textile. J.M. Szymanski uses forging techniques combined with modern machine tools to create a unique, carved iron frame. Neoset, a digital design studio, carves the stone using computer and robot technology. Modern production techniques combined with classical materials and design elements make this a unique and very special sculpture.

Textile designer Alexandra Kohl and furniture designer J.M. Szymanski met in the Hudson River valley of New York in the fall of 2014. A respect and fascination with each other's work quickly led to a professional design relationship and close friendship. Together, they develop and explore unique one-of-a kind designs that take on dramatic forms made from horse hair, textile, and blackened and waxed iron.

Alexandra Kohl was raised in New York and studied fine art at Skidmore College. She currently lives in Brooklyn New York where she runs her design practice. J.M. lives in New York City where he designs and fabricates his work. He studied at The Fashion Institute of Technology and at The School of Visual Arts in New York City. He went on to design for William Sofield.

SUZANNE TICK, MARY WALLIS & RICHARD ROEPNACK

This mockup is a prototype for a two-story atrium in a corporate office. As an artist and textile weaver, Suzanne was enlisted to create a woven sculpture for their reception lobby. There was a four-month window for design, implementation and install, the art piece had to be equal parts craftsmanship and technology while mirroring the philosophy of the corporation.

This piece for Venice Design represents the method behind creating Woven Neon, installed April 15, 2019. Process has always been the foundation of Suzanne's work ethos. Whether it's designing textiles, floor covering or neon lighting, she begins with fiber and structure. The interaction with individual materials and working through each detail is imperative to create a purposeful outcome.

The simplicity of the materials drove Suzanne to create a sculpture based on weaving light. The three materials used are neon, silicone and aluminum. The progression began with what warp and weft material to use and how the neon would be held and housed. Four weave structures and two silicone cord sizes were considered. Ten neon colors and five striae color rotations were developed. Six frame finishes, and four frame corners were fabricated. For the frame, eight color finishes were studied. The warp and weft are made out of non-conductive silicone material, and wire core silicone is used for connecting the ten colorful strands of neon. This frame represents the ten strands used for the mockup. A scaled rendering for the neon fabricator with bend details. A jpeg rendering of one of the five color rotations with weaving details.

Teaming up with Mary Wallis was key for bringing this concept to fruition. Mary is a master at working through the nuances of neon and lighting design. The third part of this equation is Richard Roepnack of Roughhouse GV LLC. Richard brings his detailed craftsmanship and skills in frame building and construction. Suzanne and Richard have been working on projects together for over two decades. The omplex frame of *Woven Neon* would not have been possible without the help of master metal design/fabricator Steve Cardellini.

CHANTAL TRAMASURE

My work has the interplay between traditional forms and sculptural expression. I use the vessel form because vessels have an interior and create a mysterious connection between the inner and outer. The relief has a dynamic movement that leads the eye into the inside and the outside and opens a dialogue between the vessel's outer planes and inner spaces. The light is reflected in places through the very thin wall.

RITU VARUNI
E'THAAN DESIGN STUDIO

'NAGA ROOTS'
HANDCRAFTED DESIGNS IN BAMBOO AND WOOD

The sound of the log drum
Filtering through the firelight
Lying in the shadows of the past
Old mithun(bison) heads in array
Smoke blackened bamboo rafters
Woven red and black lines
With roots like the old banyan tree
Deep, long and strong

Design is my language. It tells my story. The journeys I've taken, the trees I've climbed and the path I'm on. It tells you what I've experienced, how I think and what matters to me. The tug of the mountain wind, the rough texture of tree bark, the play of light in a spider's web, the wave patterns on beach sand, the changing hues of the mountains at sunset, the dewdrop clinging to the grass are some of the thousands of visual and physical experiences that remain embedded in my consciousness.

The process of creation that is Design, communicates through colour, texture, material, form. It is difficult to lay down a precise beginning to a creative process because perhaps there is none. It is part of an endless cycle of evolution, conditioned by factors of influence that touch and leave an imprint. When these penetrate within, their echo is heard in design.

A life experience can sometimes forge a deep, undefinable connection, changing and influencing the path that one takes. Naga Roots is the story of my 18 year old design journey that began in 1991 in Nagaland in India, which inspired and taught me a great deal, besides defining my craft design identity. It is a design showcase with a cultural imprint, born of my own acquired Naga roots.

In India, there are ancient tribal cultures like those of Nagaland, that contribute significantly to its multicultural identity. Nagaland boasts of a rich art and craft tradition, with roots in a lost animistic past. The bison head cutlery stand, hornbill head platter, the butter knife fitted to the butter dish like a machete on a belt are products with ethnic history. The divergent expressions of these symbols may be my own renditions, but their spirit is eternal, born of a human connection that is timeless.

The designs are born from my deep love for bamboo and wood as wonderful, organic materials of diverse use. They move with the seasons and their natural imperfections are beautiful, so why hide them. So knots, fissures and ever changing grain and tone remain. As reclaimed wood and bamboo which has size constraints is used to fabricate the designs, they are based on combinations of small components. The flatness of wood and the curvature of bamboo combine symbiotically using their intrinsic properties.

I set up my design studio E'thaan in 2002. Traditional wood and bamboo craft skills remain at the core of my design work. A path that belongs to today yet retains the warmth that comes from the embers of an old fire.

ARTISAN TEAM
Khrolo Naro, Mohammad Kamil, Rajaram, Ruovi, Omwang Konyak, Nungshi Renba

WEARABLE MEDIA

The future of fashion is interactive. We design immersive, interactive apparels that illustrate a world of futuristic fashion. Our studio innovates at the intersection of design, technology, and fashion, creating cutting-edge smart apparels that senses the world around you.

For Venice Design 2019, we are showcasing our latest sound interactive collection, dedicated for music lovers. This interactive capsule collection includes a sporty lightweight coat, a short bomber-inspired jacket, and a sleeveless top. Each garment responds and illuminates when it senses sound around you. Our custom designed electronic textile, which includes a microcontroller, sound sensor, and bluetooth connectivity, is seamlessly integrated into each garment. It is modular and easily detachable from the garment. A custom mobile app is currently under development, which will enable the wearer to customize the color of the illumination, as well as how it animates to the sound. The garments function as an interface to enhance, enchant, and resonate your experience with sound and music.

Our studio practices a sustainable production cycle, including reusing upcycle fabrics to create our collections. Each garment is one of a kind, with each fabric selection unique to the piece. Our collection is available for rental or purchase. And if you are interested in integrating our modular e-textile with your design, please contact us.

WORKTECHT

BIOMBO - A MOMENT IN A FLEETING WORLD

Our piece will be presented through the application of photographic images and lighting on a BIOMBO, a folding screen made from joined panels and usually decorated with art. We have chosen the BIOMBO as the format to present our idea because we believe it is an interesting tool with its own characteristics and special meaning within interior space. Considering these features, we have defined this piece as a concept installation of an environmental lighting fixture which takes advantage of its Eastern cultural inheritance.

The BIOMBO was invented in China over 1,500 years ago. Pronounced byobu in Japanese, it literally means "protection from wind," which suggests that the original function was to wall people off from the surroundings. In time, this function merged with the Buddhist concept of SIMA, a boundary forming a sacred precinct, and eventually the BIOMBO became a movable partition to separate a singular world into plural worlds, dividing people into different groups of higher or lower status.

In addition, the BIOMBO used to have a deep relationship with life. When a newborn was arriving, a BIOMBO was used to wall off the childbirth space. This new BIOMBO was often made in white and gold to welcome the new life. When someone passed away it was the custom to turn the BIOMBO upside down behind the dead person's body while he or she stayed at home. The BIOMBO also served as a partition between the living and the dead.

For the visual art, we collaborated with the Japanese visual artist, Namiko Kitaura, who was involved with and in charge of the art direction as well. As a team, we decided to use a dragon for the catch image of the BIOMBO. The dragon is a sacred animal used only for the aristocracy in the East, and it has always been a sign of good fortune. With its incredible power, it is the best choice for the image to apply to the BIOMBO to fit the idea of sacred precinct. An abstract style was used for the photographs to express the movement and skin of the dragon.

These dragons are the medium through which we connect our present time with that of our ancestors by means of the same figure. Photography is the modern way to record our existence and memories. We believe that our lives are many independent moments that can exist in parallel. That is also why we wished to develop our creation with inspiration from ancient times.

Finally, we wish to thank the members of our community, LOOG: Lighting Sou Co. Ltd. for the spotlight, Luci Pte. Ltd. for the indirect light, and Rayos Ltd. for control system design. We also appreciate partners Shimodaira Co. Ltd. for manufacturing the BIOMBO, KANN DESIGNING OFFICE for developing the leather hinge, as well as LOOG members Ricardo Architectural Lighting and Asterisk for sharing the fee. Without their help and that of Namiko Kitaura, we at WORKTECHT could not have turned our design into reality.

JUNGMO YANG

As a designer, I constantly research the efficient and effective utilization of materials and manufacturing techniques with curiosity. To do so, it is crucial to develop my imagination for the well-refined objects.

I always face limited conditions, for example, to use a limited material, manufacturing technique and production system whenever starting a new project. I don't limit my extraordinary idea on the object even if it seems hard to achieve in reality. My flexible response towards pieces enables me to create the well-refined objects under the difficult circumstance.

GYEOL CHAIR
The initial intuition was that of a chair which embraces a natural texture like a wood. At that time, I worked with a Korean furniture brand that produces the multi-function furniture using plywood and proposed a new chair which can be used with other items of the furniture brand. They accepted this proposal that allowed me to develop a new chair.

In this project, it is important to select a stably supplied material and to simplify a production process. Considering this situation, we decided to use the birch plywood which is a popular DIY material with stable supply in South Korea. This material has a beautiful texture that matches the chair what I had thought.

First of all, I collected a wide range of the birch plywood in accordance with thickness. The thinner it is, the more it bends naturally. In this process, I found out the natural curve of the thinnest one (4mm) looks visually comfortable. Based on this, I immediately made a prototype with paper and solid foam by applying the natural curves to the seat and backrest. While I appealed to the manufacturer, the natural curves could improve the efficiency of production. From their point of view, it was a reasonable way to choose a handwork process in order to save cost and time rather than going through a plywood bending process.

Whereas the first sample had the form of the attached seat and backrest, it was modified to separate the seat and backrest for more efficient production.

Gyeol means a texture in Korean. Gyeol chair is made from the thinnest birch plywood and the beechwood frame. Thin plywood bends naturally, which makes the comfortable curve both visually and physically. I apply it to the seat and backrest of this chair. This naturally bent thin plywood chair is completed by the handwork of manufacturer.

YOUR ARTIST
ALEXANDER LORENZ

As a designer as well as an artist the main interests of my work are:

- lightness
- sculptural expression
- clarity (construction and technique)
- ecology

The chair "*No 6*" includes all of these ideas and as a piece of design one more important idea: ergonomics. This point is the visual motivation of it's design: the chair´s lines merge with those of the human body - the design is completed with it's user.

Worldwide it is the only cantilever chair build as a classical wooden framework. Therefore, the statics and the choice of the materials are most challenging.

But after all the technical efforts the chair "*No 6*" seems to dance and to invite You to take a seat.

To distribute my vanguard design as well as my art work I've founded the studio "*My Artist*" which organizes the development, manufacturing and selling.

The symbiosis of Art and Design in many of my works now has found a perfect roof.

MONIKA ZABEL
URBAN PILGRIMS

"How we dress is a means to express ourselves. We can send powerful messages...."

Ode to the Ocean – Water. Fishes. Fashion.

Fashion Industry, in particular fast and mass fashion production, is a major contributor to global pollution of water, air and soil throughout its lifetime cycle. Demand for clothing items worldwide has doubled since the year 2000 while the frequency of using each item before disposing it has dramatically decreased. And so has the quality of clothing. Increased use of synthetic fibers in the composition of fabrics has contributed to the plastic waste pollution and micro particle load in the water and in fishes. Researchers predict that by 2050, there will be likely more plastic in the ocean than fish. Global waters and life below water are in a critical condition.
And so is the fashion sector. Time for a fashion disrupt......

Conscious consumption of precious resources has concerned me for many years and has influenced how I live and how I design. The couture fashion collection shown at Venice Design 2019 recognizes this challenge and introduces a conscious luxury collection that is unique, of high quality in the choice of materials and how it was conceived.

As a fashion designer I am pairing beauty with sustainable design techniques (zero waste, reuse, repurpose, refashion), fine natural materials and couture tailoring. I am using resources already in the system, like end of role, samples, overproduction, stock and scraps (precious materials that sometimes are described as pre consumer "waste") and apply design techniques that use or re-use fabrics efficiently, like zero waste. The results are unique pieces of fashion art. My leitmotiv and design brand is Urban Pilgrims. They are, like me, travelling, discovering different spaces and places and changing them as well as they are changing themselves.

Ode to the Ocean is an installation that stages a couture collection consciously and ethically designed for Venice Design 2019. The four fashion models and fishes interacting in a symbiotic, light and colorful way. A small boat (an arch?) floating in a sea of shells, being overlooked by models and fishes. Shells are the symbol to guiding pilgrims on their trail to enlightenment.

Ode to the Ocean wants to raise awareness for inner beauty and valence of clothing. It calls for conscious clothing decisions at all levels – designers, producers, brands/marketers and consumers. And for respect for nature, for the element water and life below water, which is also the life line of the city of Venice.

HONGTAO ZHOU, FANG QI, ABAIYI-AKEBAI & LIANGLU HOU

TEXTSCAPE-TONTSEN EYE (SCULPTURE PIECE)

Printing was first created in ancient China as an efficient and effective means to reproduce text via the woodblock printing method. Over the centuries, the technology of printing has widely expanded to include such innovations as 3D printing, an additive process which constructs 3D objects in addition to reproducing text. Artist Hongtao Zhou has made notable innovations in 3D printing and art. He has created the Textscape, an art form that employs 3D printing technology to create a hybrid of text and sculpture, manipulating the text to form cityscapes of dense urban metropolises such as Venice, Shanghai, and New York. The text functions as legible maps, visually echoing the realistic skyline as well as describing the cities' demographic data, and calling attention to the notions of space and/or lack thereof.

This is a large-scale Textscape comprised of industrial steel representing a cityscape of Shanghai, a rapidly growing urban city in China. The text will describe the architecture history of Tontsen Architects, whose mission is to create innovative and thoughtful designs in urban planning, architecture and construction design. Together, the cityscape of Shanghai with the text comments on the rapid urban development and its effects on the skyline and surrounding natural environment. It prompts the viewer to consider the significant ramifications of such fast-paced development to the earth's ecology and sustainability, inviting the viewer to pause, consider, and reflect. The artists question what ways we may collaborate and find effective solutions to reach a balance between industrial development and ecological sustainability during this dynamic time.

CLOUD OF MEASUREMENT (HANGING PIECE)

The ruler is a tool that was initially created to measure length. Today, it has become a tool in the Chinese school system to measure scores in education for the general public. However, by doing so, the restrictive and reductive parameters have indirectly limited creativity. The Cloud of Measurement, the hanging installation displayed in the exhibition, is comprised of twisted triangular plates created by students who have recently completed their exams by using extreme heat or 'hate'. They found the process of making this piece as a cathartic means to release their emotions after their intense examinations and to express their creativity via the transformation of a rigid object into a fluid form, presenting a symbol for a nonlinear measurement for creativity. Collectively, these twisted rulers form a cloud-like shape which reflects their convoluted educational system, critiquing their school life, and prompting the need for change and action.

Together, the Eye is sailing through the ocean underneath the Cloud to search for solutions for the unknown future.

PARTICIPATING ARTISTS

Yumei QI, Kewei FENG, Jiabao ZHU, Xiaotong ZHANG, Hongliang WANG, Guoyi ZHOU, Marcos Cruz Ortiz, Wenjie NIE

MADE IN VENICE

DESIGNERS IN VENICE
BY ILARIA MARCATELLI

Made in Venice, also this year supported by the European Cultural Centre, is now an integral and consolidated part of VENICE DESIGN. The project aims to promote local craftsmanship through the selection of 42 ateliers of different specializations, each representing the highest quality and tradition of crafts in the city of Venice.

The research of *Made in Venice* is a continuous process. Since 2017, when it was born, a total of 56 workshops have been mapped. Besides the artisans we already know and their new interesting projects, this edition includes six *new entries*: Artefact Mosaic Studio, Paolo Brandolisio, Toni Dalla Venezia, Riccardo Guaraldi, Daniela Levera, and the Minelli sisters - more to be revealed in the following pages.

This is how the journey to discover Venice continues: through its *calli* (alleys) and *campi* (squares), the visitors find treasures that are becoming more and more precious.

Paradoxically, in a city that is evidently recognized as a museum, its uniqueness and authenticity are struggling to survive.
Therefore, only few courageous and passionate carry on the tradition of the craft trade, which has always been Venice's source of pride, thus giving life to extraordinary products.

In addition to the artistic value, the benefits of craftsmanship for the environment should not be disregarded. Especially nowadays, when the attention to our ecosystem is, or should be, a priority.
In line with this edition of VENICE DESIGN that showcases reflections on the use of technologies, environmental problems and well-being, *Made in Venice* will also draw the attention on the sustainable aspect of craftsmanship.

In fact, every day these craftsmen work in absolute respect for the environment, through an ethical and conscious approach to each of their disciplines.

All the choices involved in the creative process result in the durability of the product, while preserving its value and technical properties: from the selection of materials to the attention for details, and the meticulousness in all stages of the production.

Although craftsmanship is often considered an old, if not an outdated practice, it contributes to a sustainable economy by decreasing material consumption and by reducing energy and pollution for its disposal. It demonstrates a far-sighted vision and conscious attention to future well-being. Small craft businesses, in addition to creating *beauty*, are beneficial to our ecosystem.

Artisans interpret and satisfy the needs of society, by manually shaping the material, transforming and personalizing it into authentic and absolutely unique forms. They are specialists of the material they work with and know its potential and limit, starting from the tradition and experience passed from those who preceded them. Thereafter, they reinterpret and develop their ability to explore more and experiment in this material to *feel*, *touch* or *listen to* it - sometimes quite literally.

With its third edition in collaboration with VENICE DESIGN, this project shows the most diversified fields and techniques of Venetian crafts practised every day in the lagoon: spotted and reachable through the VENICE DESIGN map.

ALTROVE

ARTEFACT MOSAIC STUDIO

In this atelier of clothes, situated in the San Polo neighborhood, every idea starts from the meaning of the word Altrove, as the sensation to go beyond spatial and temporal convention.

"Altrove is a word that we always liked, maybe it is the word that best describes dreamers. But it is not a *somewhere else* interpreted as a will of escape from something. It means to be constantly somewhere else, with a strong positive significance. All of our clothes come in some way from a geometrical shape. They all are developments of forms. Forms without limits" Alessandra states.

Functionality and wearability create new volumes, fabrics are meticulously selected and all made in Italy. They seem to go with the body, creating real architectures of clothes for human beings. An aesthetic that makes an expressive use of structure, inflected in monochromatic tones and combined with a precise tailoring knowledge: it is just through the personalization of lines that it becomes contemporary.

Artefact Mosaic Studio is an innovative workshop in the field of manufacture of mosaics, a tradition dating back to ancient times. It also produces high-end Italian contemporary mosaics. The company brings a fresh and original approach to the Italian mosaic by providing a complete project management.

Their mosaics are entirely handmade at all stages, from the sketch, along design and production. The staff is only composed by professional and graduated Master Mosaic craftmen, just as owners Alessandra Di Gennaro and Romuald Mesdagh.

The respect of the tradition in a contemporary approach, the craftmanship, the creativity and the constant search for new materials and solutions make what Artefact Mosaic Studio is. They understand the new luxury of handcrafted, bespoke mosaic as the unique experience of truly customized service. From their proposals, they help each client to create its own sketch and unrepeatable high-end piece.

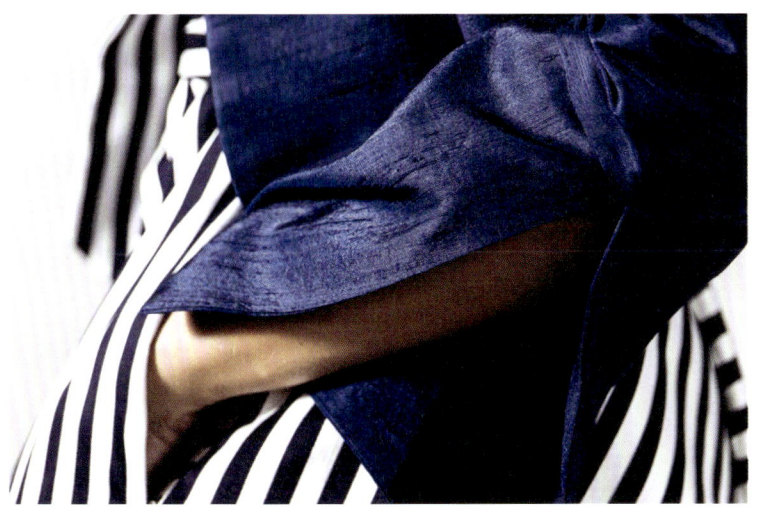 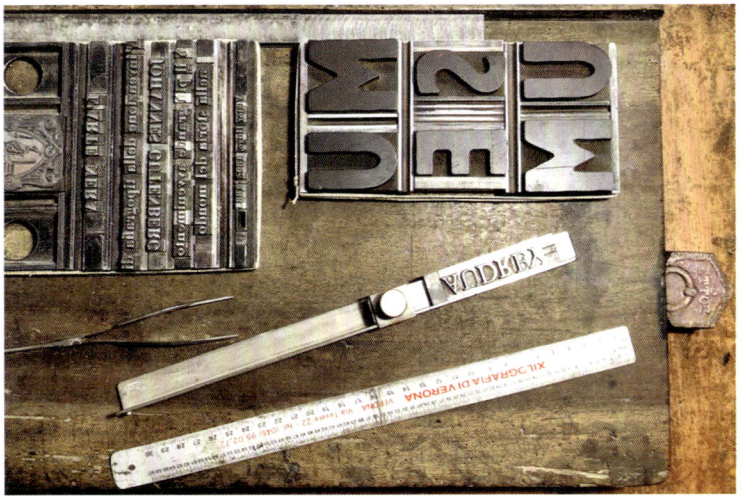

BANCOLOTTO

TIPOGRAFIA GIANNI BASSO E STEFANO BASSO

The Social Cooperative Il Cerchio has been addressing for more than 20 years to all the issues regarding the inmates and former inmates. In 2003, Il Cerchio, helped by the government which provided the facilities and the spaces, created a tailoring workshop inside the female correctional facility in Giudecca Island, Venice.

Nowadays the coordinator of the tailor's workshop teaches to the inmates the art of craft and dressmaking, in order to promote this creative and artisanal activity. All the creations are made with the finest fabrics (Fortuny, Bevilaqua, Rubelli, etc), and are sold in the shop "Banco Lotto N 10" located in Castello District of Venice. Il Cerchio is also collaborating with important Organizations (Teatro La Fenice, Giorgio Cini Foundation, Cipriani etc) in national and international projects. Since 2013, during the International Film Festival, the cooperative has the possibility to set up a temporary shop where handcrafted creations are sold. All these projects help the cooperative to increase the production and the image, thus delivering to all the prisoners a big satisfaction and a big hope for their future.

Located in a quiet street of Cannaregio, the renowned printing laboratory of Gianni Basso is difficult to find. He does not fancy modern technology. Therefore anyone who wants to get in touch with the "Gutenberg of Venice" should send him a letter or reach him on his rotary-dial phone, just as his famous clientele (including Hugh Grant and Nobel Prize winner Joseph Brodsky) would do. And it is worth it. Trained in letterpress printing by Armenian monks he is now passing on his trade to his son Stefano.

The genuine and welcoming print shop is a time machine. The bookshelves next to an 18th century press are covered with exquisite ex libris and lithographies that have captured the atmosphere of Venice as it was 150 years ago. For his designs and layouts, Gianni has an extensive collection of magnificent old woodcuts and copperplate engravings that he has painstakingly collected. Enchanting.

MARIO BERTA BATTILORO

TESSITURA LUIGI BEVILACQUA

Mario Berta Battiloro was founded in 1969, with the aim to carry on the ancient family craft begun in 1926. In the historic laboratory – a former home of the Renaissance painter Tiziano Vecellio in the sestiere Cannaregio - gold, silver and other precious metals are transformed into ultra-slim leaves. The cornerstones of the business are its artisan production and its handcrafting.

Respecting 16th century procedures and using only manual tools guarantee a product of superior quality because the raw material undergoes less manipulation. The fusion phase (eliminating the metal's impurities) and the goldbeating process (the hammering done by master Marino Menegazzo) are fascinating to watch. The foils will then be suitable for applications mostly in the art field.

They can notably embellish mosaics just like the ones of Saint Mark's Basilica or parts of gondolas. The company is extending its production to the food and cosmetic sectors which let them explore innovative possibilities.

The Tessitura Luigi Bevilacqua, led by the Bevilacqua family, carries on one of Venice's most ancient traditions, using original 18th-century looms as well as mechanical production. The velvets, brocades, damasks and satins they produce show the same quality as those of centuries ago, because they are made using the same techniques and looms, with patterns coming from various centuries and different corners of the world.

Luigi Bevilacqua moved into the current building in the Santa Croce district, where the old looms have finally found a home. The Venice premises now host part of the production facilities, too – with 25 hand-operated and still working looms – as well as the warehouse and showroom. The Bevilacquas run their company themselves, their constant presence and solid knowledge of weaving techniques guarantee a high-quality and renowned products. Their efforts to improve the brand's prestige results in strengthening trade relations in all parts of the world, though their fabrics will still be niche products.

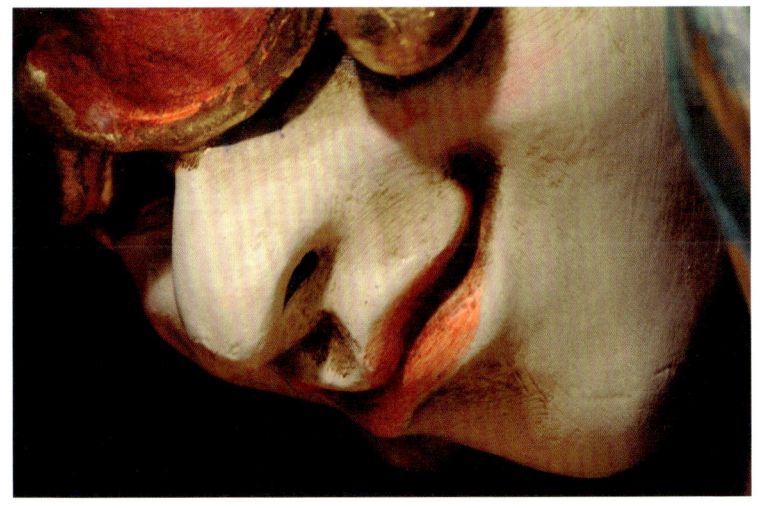

LA BOTTEGA DEI MASCARERI

BOTTEGA ORAFA ABC

The welcoming brothers, Sergio and Massimo Boldrin, have been running a shop at the foot of Rialto bridge since 1984 and making masks for 35 years. The two passionate creators also own a workshop in San Polo where you can watch them making their incredible handmade figures following centuries-old craft. Clay is used to create the shape, alabaster is then poured over to create a mold and finally papier-mâché fills it. Paint, gold leaf and lacquer inject life.

Their masks lead into a world populated with medieval protagonists: from jesters to court jongleurs, and obviously with characters of the commedia dell'arte. The collection is also remarkable for decorations drawn from Tiepolo, and original lunar and solar pieces are evocative of Venice's carnival. Often showcased in worldwide exhibitions, their masks are mainly produced for the theatre and cinema (as in Stanley Kubrick's Eyes Wide Shut) and can be purchased. The shop and atelier will leave the visitors with the feeling to have dived into original and high-quality Venice's culture.

ABC goldsmith, run by Andrea D'Agostino in San Giacomo dell'Orio since 1998, is identified by its exclusively handcrafted jewelry production, the result of a continuous research in innovative materials and of elegant and sophisticated design. These jewels are distinguished not only for the aesthetic taste, but also for the harmony they create between their shape and their extremely comfortable wearability.

ABC goldsmith produces jewels in mokume-gane, an ancient technique and unique in Italy. Wonderful patterns are born combining different metals: silver, copper, yellow, white and red gold, the combinations can be almost infinite. The result you get is the uniqueness of jewelry, never identical, individual, which makes it even more exclusive. Each jewel can be customized with the technique of mokume-gane, that embellishes the jewel, and, thanks to a long and laborious process, grants a unique design. Andrea can be seen at work in his atelier behind the shop.

PAOLO BRANDOLISIO

ANNA CAMPAGNARI

It is rare that an artisan's concern, other than the scrupulous execution of work, consists also in creating harmonious and pure sculpture. But hidden amongst Venetian alleys, behind San Marco square, there is one of the few remaining oarlock artisans.

The workshop, which was in the past owned by the oarlock master, Giuseppe Carli, has now been handed down to his apprentice, Paolo Brandolisio. It was with trepidation that Carli taught this old tradition to his pupil, knowing that although the craft could be acquired over time, the artistic ability was an inmate, unattainable quality. That was 35 years ago, and Paolo still remembers when he was waiting his master outside the workshop.

At the age of 51, Paolo continues searching new ways of evolving his concept of the perfect oarlock, as always conceived as a sculpture. He has adapted his teacher traditional methods to his innovative style in order to address modern issues related to the lagoon. In Paolo's altruistic search for the balance of art and design, it is possible to perceive the oarlocks feminine lines evoking primordial forms.

Considered one of the top Venetian rowers, Anna's long career as athlete includes four wins in the Regata Storica, the biggest traditional event on the Venetian calendar. In 2009 Anna opened her own artistic sewing laboratory to produce the traditional flags awarded to the winners in Venetian regattas. These pennants are completely handmade for each race and date back to medieval times. They are hand painted with illustrations regarding the festival the regatta celebrates and are tied to wooden staff decorated with gold leaf, also produced in her workshop.

Anna's work over time expanded to include other types of traditional flags and banners. She continues to express her artistic craftsmanship with a number of original pieces which can be viewed in her laboratory. This delightful workshop on the ground floor of her family residence is open to visitors by appointment and gives a rare glimpse into a traditional Venetian villa. Anna is very pleased to share the experience and techniques regarding her craft works. A unique insight into the traditional Venetian world of boat regattas and all the crafts around them.

CANESTRELLI

CARTAVENEZIA

Ancient techniques to create modern objects of furniture: this is what inspires Stefano Coluccio. After obtaining his degree in architecture in 1996, he decided to pursue his family's artisanal tradition, started by his maternal grandfather, the engraver Emilio Canestrelli, and followed by his mother, Manuela. Today, in his shop-laboratory in the heart of Venice, close to the Accademia Gallery, he brings to life sophisticated and elegant mirrors. His inexhaustible source of inspiration is art history. Especially in the paintings of the most celebrated Flemish artists, e.g. Jan van Eyck and Quentin Matsys, or Italians like Parmigianino, Bellini and Caravaggio, these witch's mirrors, known also as sorcière, were depicted frequently.

All the mirrors are designed by Stefano Coluccio and produced in his workshops in Venice. Using an artisanal process, the Italian designer makes only unique pieces, which are the result of a constant research and experimentation with shapes, ideas, and design.

The contemporary paper artist Fernando Masone was born in 1952 in Pietrelcina, Benevento, Italy. In his twenties he discovered art in Rome while working at the art studio Esedra before he attended the Scuola Internazionale di Grafica in Venice. In 1980, he started in Rome with ceramics studies to finally specialize himself in art print. Today, Fernando Masone has his own laboratory and is organizing workshops in Italy and abroad. In 1990, an expert in modelled art print and special handmade paper, he opened a laboratory of handmade paper in Giudecca.

Paper designer as well as hand papermaker, Masone conceives his own creations and collaborates with contemporary artists and makers of books. Cartavenezia is located in the dynamic and creative "Chiostro Santi Cosma e Damiano". It is a gallery, an art shop and a workshop where Fernando Masone cooperates with artists and showcases his work.

MARISA CONVENTO

I VETRI D'ARTE DI VITTORIO COSTANTINI

She is an Impiraressa, a beadstringer: this is the original name of the ancient Venetian craft practised in the past by so many women in Venice. The seedbeads and beads produced in Murano and Venice needed to be strung, threading them with very long needles, in big bundles so that their packing, shipping and trading would be easier. The beads then would take long journeys to the far continents where they were highly desired by the native people of Africa, America and Asia. The traditional techniques of the Impiraresse and the precious vintage seedbeads, as little as a pin head, are at the base of her work.

Bigger and intricate lampworked beads, famous all over the world for their beauty, and used in the past as money for the trades, are made for her by the best contemporary bead-makers. She creates necklaces, beaded flowers, corals and embroidery at her own design and imagination, never repeating the same piece.

Her way to respect the heritage of her ancestors, a tribute to Venice, the city where she lives and works.

Vittorio Costantini was born in 1944 in Burano. He began an apprenticeship in a glass factory at the early age of 11. Since he opened his own workshop in the Castello neighborhood in 1974, flame-working has become his only true focus. He always had an innate fascination for nature and all his creations show great mastery and passion for it. He spends endless hours creating individual pieces: from multi-colored insects to iridescent butterflies, birds, fish and flowers. All as the result of his manual skills.

Vittorio considers himself a great observer of the microcosm we are surrounded by. With the profound vision of an artist, he can see deep into the fields, the skies, the waters. His artistic talent has led him to participate in numerous exhibitions in Italy and abroad. In the past few years, he has enjoyed devoting himself to teaching and demonstrating. His rich, personal collection inside the workshop is the testimony of many years of flame-working and the evolution of his technique.

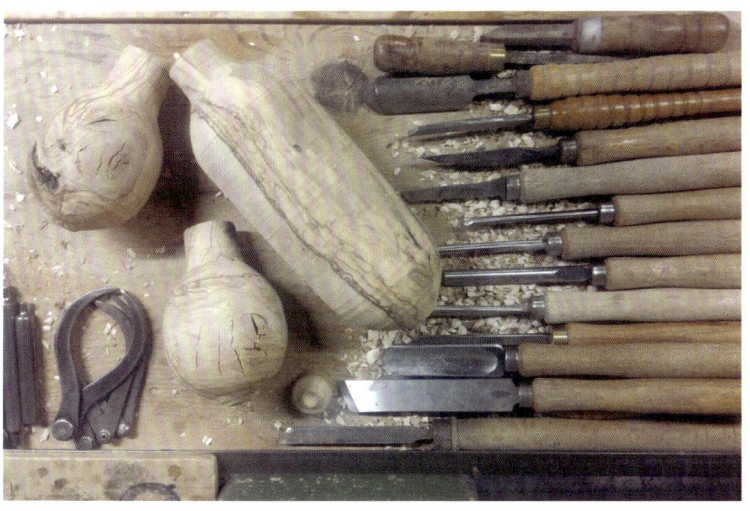 

ANGELO DALLA VENEZIA

TONI DALLA VENEZIA

Angelo dalla Venezia represents the last turner in Venice. In 1959 he began working with a wood lathe and some tools given by his previous employer, the master Vio Vincenzo, and he still continues to this very day in his workshop, a few steps away from Campo San Polo. With the advent of mass-production, the demand for custom orders decreased dramatically and the number of turners consequently plummeted. Driven by the love for his work and extremely determined to carry on his activity, Dalla Venezia began to create also some decorative objects alongside the more traditional utilitarian items. By reinterpreting an art tied to the past and local traditions, he managed to give birth to a laboratory in which you would find high quality design objects.

Today, he produces from wooden spheres not only eggs, fruits and spins, but also rings, pencil-holders, knobs and pendulum clocks. All these objects are made by Dalla Venezia with a manual lathe and different kinds of wood, but above all with the passion for his work.

Toni Dalla Venezia is an art framer, gilder and restorer. When very young, he acquired the methods and instruments of his work as an apprentice of the old Venetian masters. From 1958 to 1973 he has been living and working in Cannes. Here he could develop his preparation and activity in contact with the artistic contemporary milieu acting in Côte d'Azur. He worked for Sir Douglas Cooper and framed many paintings of his collection. Afterwards he personally knew and worked for many important artists like Picasso, Ozenfant and particularly for Graham Sutherland with whom he maintained friendly and professional relations even when he returned to live and work in his native town, Venice. Toni Dalla Venezia's frames are entirely conceived and made by him: from carpentry to intermediate phases like preparation of plaster, decoration, gilding to the final lacquering. Since 1973, the workshop of Toni Dalla Venezia is in Venice, and since a few years in Cannaregio 6021, close to the church of Santa Maria dei Miracoli.

DECLARE

DOPPIOFONDO

Declare is a leather brand-au-courant giving new meaning to local handcrafted production and design in Italy. Situated in San Polo neighborhood, the contemporary showroom of Declare proposes singular bags with a design aesthetic inspired by architecture and couture backgrounds. With an eye for detail and function the Declare design is sealing each item with their signature merge of the highest in quality and 'of-the-now' in design. Offspring of the rapidly moving fashion world that breeds them, co-founders Omar Pavanello and Emanuel Cestaro seamlessly channel enduring style standards with a nod to the glitz of luxury cool. Their creations are colorful flashy pieces combined with sharply classic volumes where seductive serpentine textures reflect playfulness in the functional and every day to night. Into the Venetian store, a newsstand fills up an entire wall with the best independent magazines of the world, this is called MAGWALL, a project run by Declare together with EDICOLA518.

DoppioFondo arts and culture no profit association was founded in Venice in 2011. As a fine art print studio and independent publishing house, it is specialized in organizing workshops, artist-in-residence programs and art projects. The non-toxic printmaking studio is fully equipped for work in etching, engraving, woodcutting and silk-screen printing. *Edizioni DoppioFondo*, among others, is one of its main projects and it aims to support the artists in the realization of their own book and self-made printed material, interpreting them as the expressions of a personal work or as the result of an artistic partnership and creative process.

The goal in this project is to promote books in a contemporary context, highlighting the importance of tradition-inspired printmaking techniques. This is the reason why DoppioFondo choose to handprint the books in limited edition and using tradition-inspired. Inside the laboratory, there is a small space where you can find prints, books and other original stuff self-made originally DoppioFondo.

EMILIA BURANO

IL FORCOLAIO MATTO

Lorenzo Ammendola was born in 1970 on the island of Burano. He grew up in the studio of "Emilia Burano" (the mother's name of his great-grandmother) which for four generations creates the original lace of Burano. Ammendola reviews the history, the pageantry and the passion of his family that in all these centuries kept creating and brought this ancient heritage to the present day. He then renews and searches for new forms and ideas. The study of the many similarities of Venetian lace with decorations and architecture of the buildings found its origin in the 90's in the collections inspired by the most beautiful palaces in Venice. Of importance is the study and the realization of the first sculptures made of Burano lace with the same techniques of the 1500's.

Lorenzo is an eclectic and passionate "designer in motion" and his design studio is also involved in automotive. Collaborations with world renowned brands such as Aston Martin, Rolls Royce and fashion designers helped to increase his knowledge and experience worldwide.

Just off Strada Nova, the main street of Venice, the workshop of Il Forcolaio Matto is somewhat hidden. Master Piero Dri is the youngest *remér* in Venice, making oars and *forcole*. The smell of wood and the taste of traditional Venetian craftsmanship create a warm and welcoming atmosphere. Born in Venice and graduated in astronomy, Piero chose to dedicate his life to his passion for Venice and rowing back in 2006.

Carrying on a centenary tradition, dating from 1307, he learned the art of making a *forcola* from his master. The *forcola* is the crutch of Venetian boats, the base on which the gondolier places his oar to steer the gondola. Custom made for each oarsman, *forcole* take on special sculptural and dynamic lines, as a result of a thousand years of history in constant research of a perfect balance between function and beauty. As essential tools for propelling the gondolas through the canals, *forcole* became both the symbol of Venice and an art piece highly appreciated all over the world.

FORTUNY

ATELIER SEGALIN DI DANIELA GHEZZO

More than a century old, Fortuny remains the highly esteemed Venetian textile company founded by artist, inventor and fashion designer, Mariano Fortuny. Under management of the Riad family for nearly 30 years, Fortuny continues to be infused with the spirit of its founder. Every fabric is still produced in the same factory on the island Giudecca, on the same machines, using the same process and techniques as developed by Mariano Fortuny over a century ago.

Just as he combined his respect for tradition and the past with his love of innovation to inspire his creativity, Fortuny continues to be a pioneer in the world of design and technology today. The brand also produced amazing chandeliers still based on his drawings, mosaics and very elegant Art Deco furniture. The pieces are all modern re-interpretations of Venetian classics. Although the factory itself allows no visitors in order to safeguard trade secrets, the showroom conveys a warm atmosphere. The adjacent gardens can be visited by appointment.

Segalin tailored shoemaking was founded in 1932 by Antonio Segalin, between the two World Wars. His elder son Rolando became his successor and worked according to his father's teaching. Daniela Ghezzo, who had worked for shoe manufacturer Gatto in Rome, expert in man tailored shoemaking, took over the family empire in 2000. Her education, based on the famous Academy of Arts in Venice, combined with her hard daily work enabled her to continue and improve the entire production of artistic, hand-crafted shoes in Venice. The studio represents the continuity of an art and an ancient job that finds new nourishment and inspiration in the daily quality enhancement and in the fulfillment of every specific wish of the customers.

Of the approximately 250 models handcrafted every year, some of the most interesting creations are on display in the window of the workshop. Her shoes with unusual shapes, original colors and made of soft and shiny leather are internationally renowned.

GABRIELE GMEINER

RICCARDO GUARALDI

Gabriele Gmeiner is working in Venice since 2003 in her workshop at Campiello del Sol, where she produces custom-made shoes of the finest quality. She studied in London at Cordwainers College specializing in the traditional shoe craft and in Paris at the Centre Formation Technologique Grégoire, for saddlery. Besides a traditional education she made a few sidesteps into the field of art.

One of her artistic projects brought her to Tokyo where she experimented with traditional crafts and materials of the Japanese culture. These works have been exhibited in Gallery ef in Tokyo and in the Historical Museum of Vienna. She has held art workshops for children and college students, and she has taught shoemaking at the Venice Santa Maria Maggiore prison in a rehabilitation project.. Her present work combines the artistic spirit and the best traditions of craftsmanship. Young apprentices from all over the world help in the production and learn the secrets of the art at her workshop.

In the last few years, the workshop of the luthier Riccardo Guaraldi is located in a really secret place: Corte Botera, a "Corte Sconta" (hidden courtyard) which recalls Pratt set in the Castello district. Here stringed instruments played in the most important Italian and European theatres are shaped.

During his life, his family has arisen in him the love of music and manual wood work, bringing him to realize his passion for violin making. It is no coincidence that Guaraldi, Venetian by birth, became a luthier in his hometown. Venice was home of some of the most important masters of violins like Matteo Goffriller, Domenico Montagnana and Pietro Guarneri who lived and worked here in the golden age of the eighteenth century, and whose instruments are now sold at international auctions for millions of euros.

The Venetian luthier never loses the attachment and the connection with his creations: they might come back to his workshop for a check, or he may run into his instruments, born from maple and spruce wood in the little Corte Botera but now playing in theatres all over Europe.

LABERINTHO

DANIELA LEVERA

Gold and silver: thanks to artistic chemistry these noble metals get a new identity in the creative shop founded in 1994 by Marco Venier and Davide Visentin.

Harmony and shapes, usefulness and completeness blended into the meticulous search for a new standard of beauty based on the combination of different artistic genres. Looking for a different aesthetic concept, materials are selected and combined: ancient seals and geometric shapes, blown glass and diamond, ebony, amber and turquoise, coral and black agate fossil, ancient stone seals, carnelian and lapis lazuli, joining in a dialogue that weaves ages and cultures. With the help of various techniques such as cantilever mosaic sculpture they create a connection that crosses ages and traditions. This is a fascinating journey into a new harmony of contemporary jewellery.

In one of the most unchanged areas in Venice is the hidden ceramic ADLCeramics workshop, founded in 2011 by ceramist, designer and ceramic teacher Daniela Levera. The workshop, in Campo San Pietro di Castello, is home to the production of artisan pottery pieces using a wide range of materials and techniques; This small but busy atelier is a meeting place for artists, architects, designers and ceramic lovers who, attracted by the material, find a place dedicated to ceramics and knowledge exchange. Daniela's personal production is deeply inspired by the landscapes and colours from her native land, Chile, and those subtle contrasts discovered in her journey - a spectacular worldwide search for colours and shapes. Her production mainly consists of furniture, household and kitchenware items. Together with Kirumakata by Alessandra Gardin, designer of glass jewels, she opened a shop in May 2018 in the area of the Venice Biennale, another characteristic place of the city.

GIULIANA LONGO

ALESSANDRO MERLIN CERAMICHE

The studio of Giuliana Longo exists since 1901, and the shop even preserves the original interior of that time. For this reason, the Veneto Region has recognized the "Local Historical Veneto" decision to protect and preserve the interior, including all shelves.

The quality of materials and the constant search for beauty, allows the skilled hands of Giuliana to create little masterpieces. It is in her studio that hats for Venetian *gondolieri* are produced as well as magical fantasies for one of the oldest carnivals in the world. Giuliana is also famous for the Panama hats that she selects personally every year and imports to Venice, almost in an extraordinary exchange of artistic craftsmanship, which links two worlds geographically very far from each other but similar in the extraordinary value of the production of special hats. In Calle dell'Ovo of St. Mark neighborhood, the craftsmanship tradition goes together hand in hand with the contemporary and an avant-garde research.

Located in the Arsenal neighborhood, there is an artistic studio, a small shop-cum-atelier, where the artist Alessandro Merlin draws his inspirations on unique pieces of ceramic. Merlin is not Venetian by birth, but according to the Dutch expert John Sillevis "Alessandro belongs to Venice and Venice belongs to him", this city is for him the perfect ambiance for inspiration.

Merlin started to draw at an early age and could, with a clear outline hand drawing, bring out his fantasy. He found his style, influenced by Jean Cocteau, Audrey Beardsley and illustrators such as Ugo Pratt and Tom of Finland. Encouraged by a friend, he started to draw and etch on ceramic and this surface became his own communicative medium. On Merlin's unique dishes appear stylized animals, Arabic mosaics patterns, seductive nymphs and his famous naked horsemen. Alessandro Merlin is now an independent artist admired by his collectors. He shares his imagination with curious people walking through the narrow alleys of Castello, who remain fascinated by his work.

 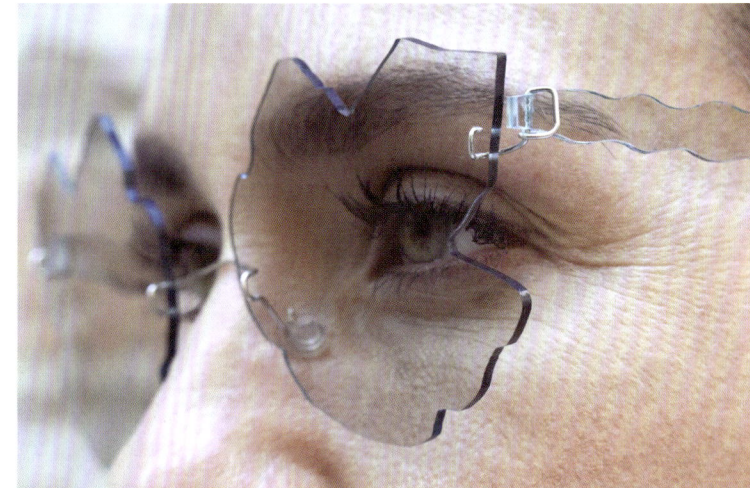

MASSIMO MICHELUZZI

MICROMEGA

Massimo Micheluzzi uses traditional techniques in his constant contemporary aesthetic research, a research that varies within almost all techniques, working with the classical ones such as the "murrino", the mosaic and the carved. The latter, offer a range from fine superficial to excavated, sculptural results, yet maintaining a disciplined vision.

Translucent pieces in fluid, organic shapes and then the contrary for more composed shapes, of absolute simplicity, often with a mono or biochromatic colour palette or in a multi-variated colour scheme, similar to the terrazzo veneziano technique with gold and semi-precious stones in opaque glassworks. Objects born and connected with the surrounding lagoon islands, delicately curved surface-carvings that echo vibrations and water-movement of the Venetian lagoon, softness in contrast to a rigid medium.

Micheluzzi maintains a rare control of his work-in-progress process together with his maestro vetraio. The mosaic panels, that the artist composes and fuses in his studio, are elaborated in the furnace by the maestro soffiatore in order to give to the initial graphic project its final shape.

Since 2000 Micromega laboratory works on the design and production of the most exclusive spectacles in the world sold in the one and only store in Venice. The idea was born from an intuition of Roberto Carlon: a special assembly system, which can be realized only through particular craftsmanship. Micromega spectacles are essential, elegant and refined.

Irreverent, discrete or almost invisible, they are known to be the most lightweight in the world. An infinite number of possible styles can be customized for each client. No glue, screws and no welding are used in these incredibly resistant frames. A wide range of materials are employed and combined, both precious and technological. The spectacles can be enriched, turning them into jewels by using gold or stones. Each client can choose any detail turning his spectacle into a unique piece. Various international patents are at the origin of this extraordinary product. Working by subtraction is the purpose, to realize spectacles made of next to nothing.

SORELLE MINELLI

STEFANO MORASSO STUDIO

Aglaia and Adriana Minelli creatively and precisely hand-print velvet fabrics. One of the most important phases of their work is the preparation of the matrix, during which the sisters decide the drawing and calculate proportions, so that in the modular print the figures can perfectly overlap. Later, they put the drawing on the linoleum and engrave it, scraping out the parts that won't be on the drawing.

The colour is prepared on a palette, the mould is passed over the colour and then on the fabric, where it is beaten with a hammer with regular and well distributed strokes. This last phase is crucial, since it is necessary to accurately measure out the colours and the pressure on the fabric in order to obtain a uniform print.

From the elaboration of an inspiration, that then becomes the idea, the project and then the drawing up to the creation of a mould and the actual press: the work of the Minelli sisters is a synergy of meticulous operations that create mesmerizing results.

Stefano Morasso was born on the island of Murano in 1962. Already at a young age, his unique talent in the *a lume* glass processing (over a burner's flame) became apparent. His natural gift for combining and matching colors was recognized early on and his innovations have been adopted and imitated by glass makers around the world.

Mr. Morasso's unique and striking style is characterized by a strong imagination, a constant artistic research, always being at the forefront of the glass processing techniques. His laboratory is an extremely creative and spontaneous place. It is situated in a cloister that used to be the convent of the Benedictine monastic complex of Saints Cosmas and Damian. Now it is a cultural location, very impressive from an architectural point of view, where he shares his passion for the craft with eight other artists and artisans with various specialties. With them he created the association Artisti Artigiani Del Chiostro.

MURANERO

NICOLAO ATELIER

Born in Dakar, the Senegalese artist (painter and musician) Moulaye Niang studied at the International Murano Glass School on the island of Murano and became glass beads maker by developing his art and by meeting Muranese masters like Pino Signoretto and Davide Salvadore.

Moulaye considers the matter of Murano Glass like a chemistry in which you never finish learning… and the magic fusion of colors makes out of each single bead a creation.

Deeply inspired by nature, Moulaye acts on a bead like on a canvas, telling in each layer of color a new story never told before. All his beads are one of a kind, so different and beautiful as only human beings can be.

Together with the South Tyrolean Emanuela Chimenton, designer and jewellery maker, Moulaye opened his first and successful workshop Muranero in Venice. Since 2004 they work together, sharing the most joyful adventure of doing what they believe in: art and beads in Murano Glass in Venice.

The costume workshop of Atelier Nicolao, founded in 1983, is known for exacting standards of research into materials, steeped in history and transformed in color and texture. Nicolao has worked on important occasions with other (Oscar winning) costume makers, making garments for movies including The Merchant of Venice and Pirates of the Caribbean. For lyric and drama, he has built stage costumes with wide international recognition. Still today, he is strongly engaged in research into historical events such as the famous Venetian *Regata Storica*. Nicolao teaches costume design at the Accademia di Belle Arti di Venezia and his costumes have been displayed at the Metropolitan Museum in New York and in the Museum of Fashion and Costumes at Palazzo Mocenigo in Venice.

In 2005 all came together in the heart of Venice: his costume workshop, showroom and the whole collection of costumes creating an elegant setting. Entering this magical space is fantastic for anyone wishing to experience the atmosphere of the past; especially today when those times are more or less forgotten.

PAPEROOWL

LE FÓRCOLE DI SAVERIO PASTOR

Paper art studio & shop. Design and creation of one of a kind pieces of jewellery, boxes, home decor and art miniatures. If you are looking for paper, you will find a wide selection of handmade sheets from all over the world or marbled and paste ones, hand dyed by Stefania. A choice of a dreamy colourful life, full of passion, as light as paper.

"I love to collect precious handmade paper sheets from all over the world because I believe that every kind of paper has Its own personality. I love to transform paper into quality design jewels and one of a kind works of art, created to enchant and amaze people. I love accuracy taking care of all the steps of the workmanship personally, from the design to the realization of the finished object, in the belief that the small details make the difference. I truly believe this is the best way to share my passion with you. My motto is: The most unexpected things reveal your personality! So come and choose the object that will bring out your style".

The *fórcola* is designed to satisfy the demanding and practical requirements of Venetian rowing, but it is also recognized as an object of art. Displayed in the most important museums in the world, such as the Metropolitan Museum of Modern Art in New York, it is universally appreciated for the beauty of its fluid, curving form.

Both the practical and aesthetic aspects of this craft are perfectly expressed in the works of Saverio Pastor. Between 1975 and 1980, Pastor worked with the last of the master *remeri*, Giuseppe Carli, the "*fórcola* king", and Gino Fossetta, the "oar wizard". In 1980, he opened his own workshop before going on to restore a workspace near the Arsenale with a group of carpenters, working there until 2001. In 2002, he opened a new workshop, Le Forcole di Saverio Pastor, at San Gregorio between the church of La Madonna della Salute and the Guggenheim Collection. Here he continues to use centuries-old techniques to make oars and *fórcole* for gondolas and other typical boats of the lagoon, adapting those used for regattas to the changing trends in competitive Venetian rowing.

DAVIDE PENSO

DAVIDE SALVADORE

Davide Penso has specialized in the design of artistic jewellery in glass and in particular in the typical Venetian lampwork beads where he developed new technical solutions and ways of innovative processing. In 2000, after ten years of experience and testing technologies, he decided to pass on his knowledge. Thus began his educational path, first with private classes at his studio, then with courses at the Glass School Abate Zanetti of Murano where he is still devoted to teaching as an official instructor in lampwork technique.

Ten years of practical teaching and many international students formed him as an educator, giving him the knowledge and ability to instruct with a simple and effective methodology to master this craft. Today the training institution "Davide Penso" cooperates widely with numerous partners, among these the "Ghana Project" of UNESCO, Boston University, School of Glass Research Bolzano, Glass School Abate Zanetti and Corning Glass Museum in Corning, NY.

Davide Salvadore was born into a family of glassworkers in Murano, where he is now living although he is often travelling for his work and to get inspirations. At a very early age he began following his grandfather into Murano's furnaces, learning how to build kilns and work glass. But it is due to his skills that he became a very unique master of glass. Utilizing centuries old Venetian techniques, his work is a continuous challenge of traditions, revealing a complexity in the use of *murrine* and a personal application of traditional sculpting techniques.

The most evident inspiration of Davide Salvadore is the African culture, with its symbols, textiles and colors. He starts from there reinterpreting and elevating it even more with his own language. Davide uses mostly soft and delicate colors of the earth, enriched by strong and bright colors typical of Africa. The sinuous shapes of his works are highly expressive and give a sense of humanity and sensuality.

MARINA E SUSANNA SENT

SMALL CAPS

Marina and Susanna Sent were born in Murano to a family with strong links to traditional glass making. In 1993, however, they embarked on a new direction: designing and producing glass jewellery. During the 1980s, Susanna, an architect, frequented her father's glassworks, where she experimented with various decorative techniques and dedicated herself to renewing the company's product lines; Marina, with a technical background, joined her sister at the end of the decade.

Today the brand Marina e Susanna Sent continues as a hub of experimentation. The two designers are not only concerned with innovating the aesthetic components of glass; they also explore it from a technical point of view. Always open to new influences, they are currently creating sculptures inspired by kinetic art, as well as producing accessories such as silk toulards printed with images of their glass objects. Remaining true to their original style, their constantly changing world is populated by new and intriguing objects.

Small Caps is an atelier dealing with graphic design. It is in Calle Avogaria, not so far from Campo San Barnaba. Small Caps was started around 2012/2013, as a reaction to an excessive use of digital devices in graphic design. Using some printing techniques (screen-print as first) was the sparkle to ignite new power in the design process: now they are able to develop an idea until the final product, a 100% custom design.

Small Caps is a place where ideas take form, made of papers and carton boards, pencils and pens, printing tools and inks, print tests and color tests. They have a cult for typography, for writing, for the shape of letters, for letterpress or calligraphy, and even today they are still using old typewriters. The laboratory is focused on Poster Art, on creating artistic affiches, feeding a research in the design and in the realization process. Small Caps is a cultural association promoting poster art and handmade prints, organizing classes to learn the do-it-yourself screen-print technique, and attending to social and cultural events with live screen print sessions.

FONDERIA ARTISTICA VALESE

GIOVANNA ZANELLA

The last artistic foundry still in operation in Venice was started in 1913 by Luigi Valese. Since 2006, his art continues in the name of Carlo Semenzato, his enthusiastic collaborator for 28 years. The foundry is located close to the Madonna dell'Orto, in a charming old industrial building dated 1797 in which the patina of time is giving a nostalgic impression of the romanticism of the early 1900 factories.

It continues the traditional manufacturing process for objects in brass and bronze: incandescent molten metal is poured into molds using the sand casting method. It is possible to visit the foundry during the afternoon and experience how Carlo and his staff create objects (they have also a shop near Saint Mark's Square). Valese's main production consists of different models of the Horses of Saint Mark, several *moretti* (Moors) and the symbol of the city, the Lion of Saint Mark, ornaments for gondolas, as well as complex objects such as chandeliers, knockers, doorknobs and handles.

"As a girl, I would buy myself clothes and accessories but then transform them by adding my personal touch. The same creative drive led me to open a workshop where I used my natural dexterity to turn my ideas into items that are highly original and unique.

I started my business by making bags, hats, fashion jewellery, and foulards, experimenting with diverse and unusual materials. After a few years, I took a two-year course in dressmaking enabling me to apply the same style I used for my accessories to prêt-à-porter clothing. My customers liked what I was doing and encouraged me to take things further.

In 2000, I met a master shoemaker, well known in Venice for his extravagant designs, and who kindly agreed to teach me the intricate craft of producing handmade, made-to-measure shoes. With these new skills, I found the best way to express my creativity and to satisfy a worldwide clientele that appreciates the contemporary touch I apply to a centuries-old craft."

IMAGE CAPTIONS

Page 157 © photo Alice Busato
Page 158 left. © photo Valentina Bernardi
Page 158 right. Courtesy of Artefact Mosaic Studio
Page 159 left. © photo Massimo Peca
Page 159 right. Courtesy of Tipografia Basso
Page 160 left. Courtesy of Mario Berta Battiloro
Page 160 right. Courtesy of Archive Tessitura Bevilacqua
Page 161 left. Courtesy of La Bottega dei Mascareri
Page 161 right. Courtesy of Bottega Orafa ABC
Page 162 left. © photo Marc De Tollenaere
Page 162 right. © photo Maurizio Rossi
Page 163 left. Courtesy of Canestrelli
Page 163 right. Courtesy of Cartavenezia
Page 164 left. © photo Claudia Rossini
Page 164 right. Courtesy of I Vetri d'arte di Vittorio Costantini
Page 165 left. Courtesy of ECC
Page 165 right. © photo Andrea Avezzu
Page 166 left. Courtesy of Declare
Page 166 right. Courtesy of DoppioFondo
Page 167 left. Courtesy of Emilia Burano
Page 167 right. Courtesy of Il Forcolaio Matto
Page 168 left. Courtesy of Fortuny
Page 168 right. Courtesy of Atelier Segalin di Daniela Ghezzo
Page 169 left. © photo Alessandro Zannoni
Page 169 right. © photo Matteo De Fina
Page 170 left. Courtesy of Laberintho
Page 170 right. © photo Nicoletta Fornaro
Page 171 left. Courtesy of ECC
Pagee 171 right. © photo Giovanni Schifano
Page 172 left. © photo Katarina Rothfjell
Page 172 right. Courtesy of Micromega
Page 173 left. Courtesy of ECC
Page 173 right. Courtesy of Stefano Morasso Studio
Page 174 left. © photo Alice Busato
Page 174 right. Courtesy of Nicolao Atelier
Page 175 left. © photo Egidio Cutillo
Page 175 right. © photo Catherine Hedouine
Page 176 left. Courtesy of Davide Penso
Page 176 right. Courtesy of Davide Salvadore
Page 177 left. Courtesy of Marina e Susanna Sent
Page 177 right. Courtesy of Small Caps
Page 178 left. Courtesy of Fonderia Artistica Valese
Page 178 right. © photo Elena Bovo

CURATORS

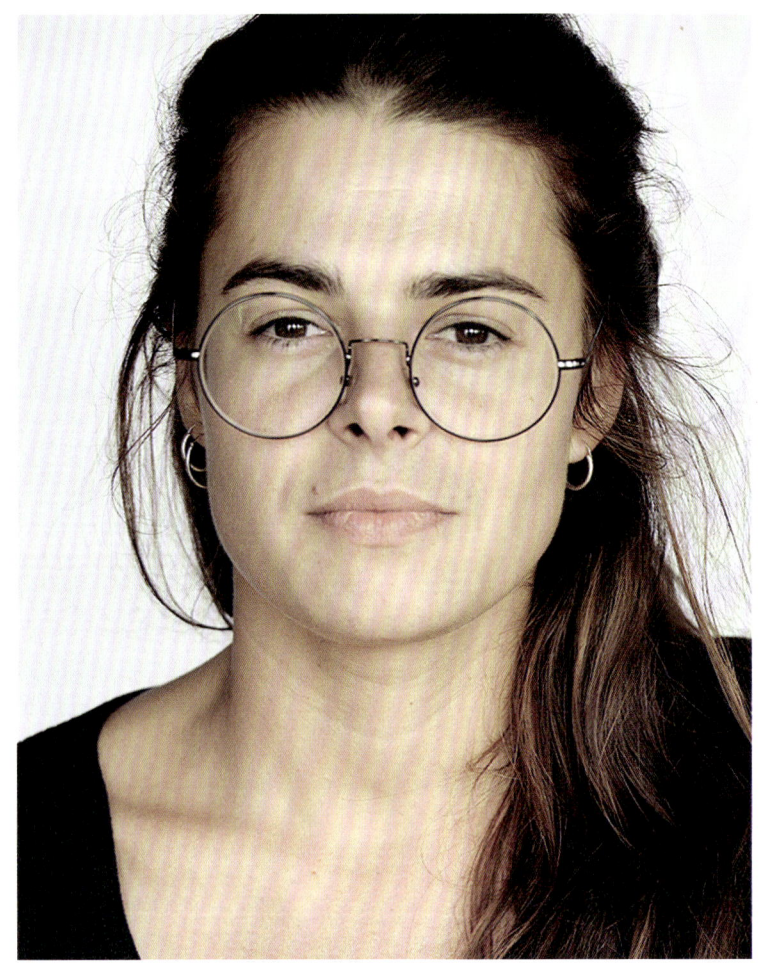

CAMILLE GUIBAUD
EXHIBITION ORGANISER & INDEPENDENT CURATOR

B.A. Media & Culture at the Design Academy Eindhoven, The Netherlands; B.A. Art & Design at ESADMM Marseille, France; and High Level technician Space Design at Condé Paris, France. First trained as designer and then as assistant curator at MAXXI Rome, Italy. Since 2017 she is exhibition organiser with the European Cultural Centre.

ANAÏS HAMMOUD
EXHIBITION ORGANISER & AUTHOR

M.A. Private Law & Intellectual Property; B.A. History of Art at the Universities of Rennes, France. Trained in Auction House and Design Gallery. She has worked as picture editor for Beaux-Arts Magazine & Le Quotidien de l'Art, Paris and on Paris International Art Fairs (FIAC, Art Paris Art Fair and Paris Photo). Since 2016, organising exhibitions with the European Cultural Centre.

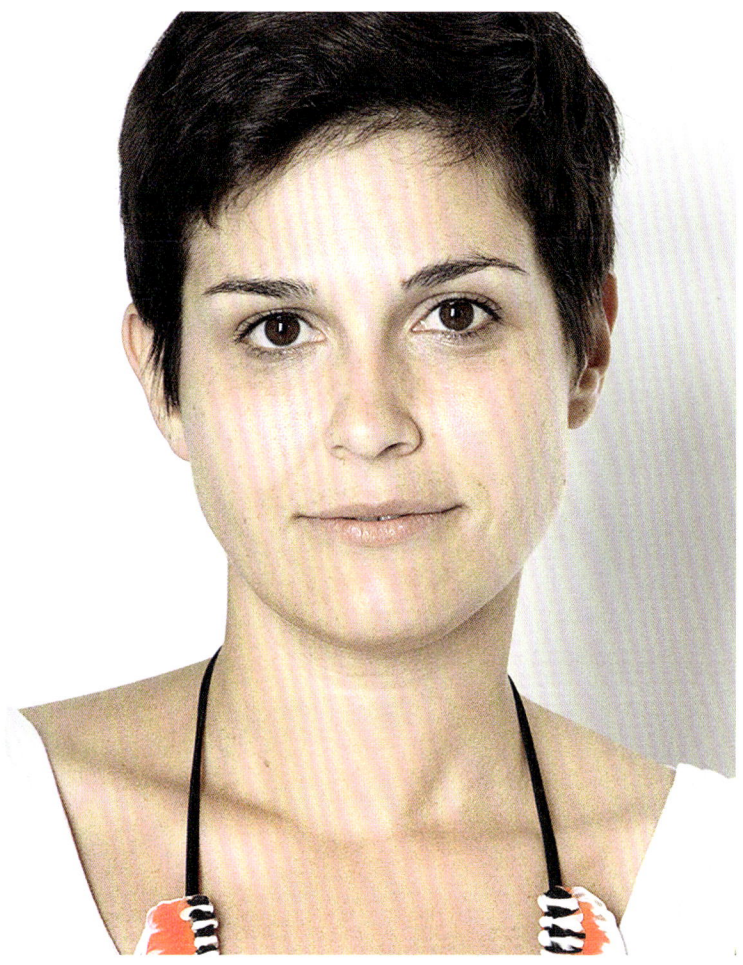

ILARIA MARCATELLI
EXHIBITION ORGANISER

Exhibition organiser and architect. M.A. in Project Cultures, Architecture at IUAV. Trained as architect at Joseto Cubilla & Asociados in Asuncion, Paraguay; Paulo Moreira in Porto, Portugal; Emergency Architecture&Human Rights in Copenhagen, Denmark. Collaboration in 2015 with the architectural event Open House Porto, Portugal. Since 2016, organising exhibitions with the European Cultural Centre.

LUCIA PEDRANA
EXHIBITION ORGANISER

B.A. Cultural Heritage and M.A. History of Contemporary Art at Ca Foscari University in Venice. M.A. in Arts Management, IED Istituto Europco di Design, Italy. Worked at the Peggy Guggenheim Collection in Venice and at La Biennale di Venezia in collaboration with national participations. Since 2014, organising exhibitions with the European Cultural Centre.

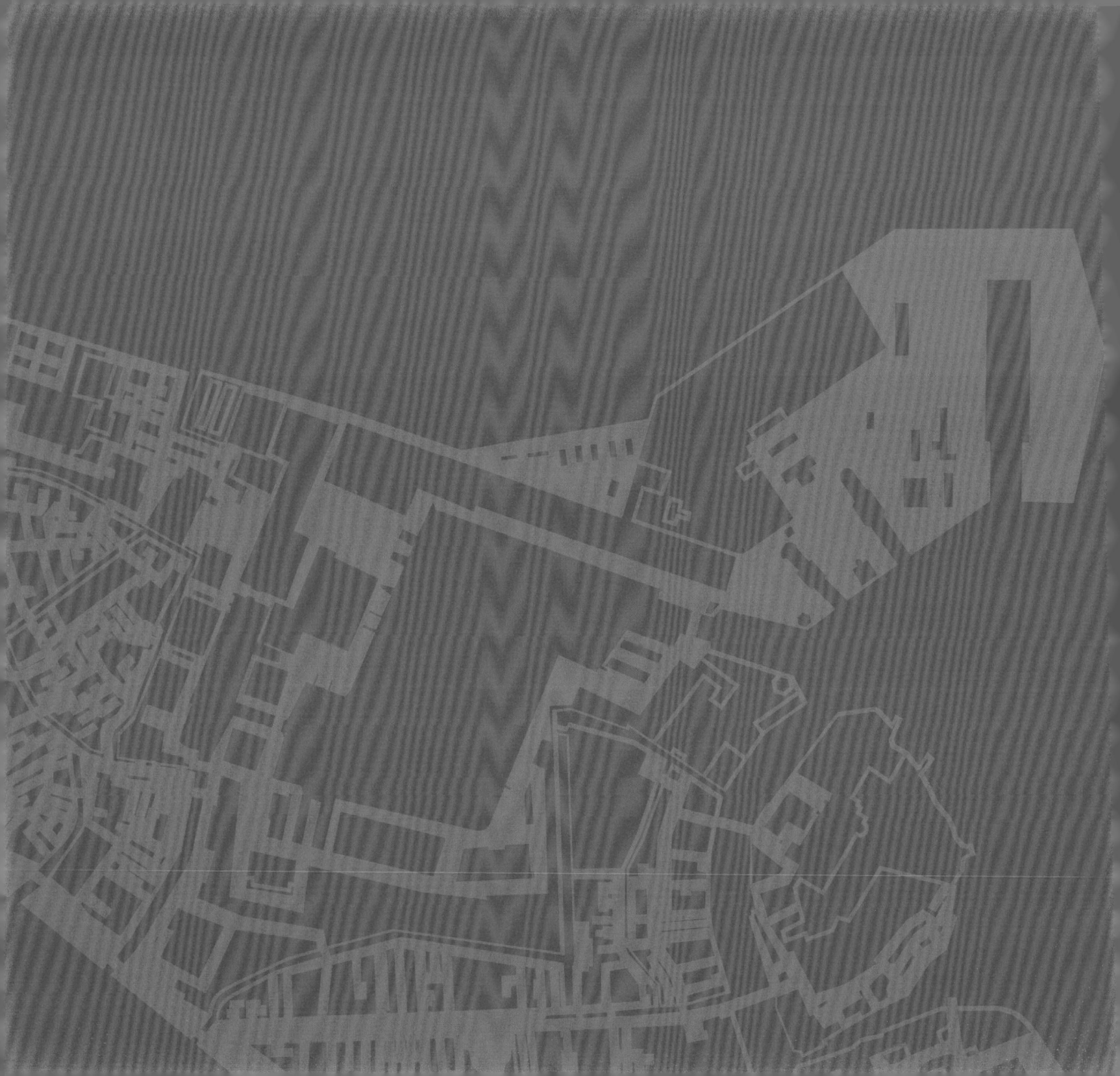

IMAGE CAPTIONS

IMAGE CAPTIONS

11	Palazzo Michiel, courtesy of the ECC Italy.
15	&SOCIETY, *Monster-L&P*, 2019. Photographer: VVORK STUDIO.
17	Farah Abdelhamid, *In-Pression*, 2019. Photographer: Natasha Yonan.
19	Andrés Aguilar, *Las Copaleras*, 2019. Photographer: Antonio Mendoza.
21	Annick L Petersen, *Alp light 310*, 2018. Photographer: Yeshen Venema.
23	Gwendolyn and Guillane Kerschbaumer, *Squares wall / floor lamp*, 2011. Courtesy of Atelier Areti.
25	Nick Boers, *Sponge table on high legs with glowing pearl*, 2019.
27	Masayo Ave, *DESIGN GYMNASTICS A.B.C. collection*, 2009-2018. Courtesy ©MasayoAve creation	Masayo Ave.
29	Alexander Bannink, *Sarco design sketch*, 2017. Courtesy of EXIT International.
30	Alexander Bannink, *3D model rendering of Sarco's front-design*, 2019. Courtesy of EXIT International.
31	Alexander Bannink, *3D rendering of rear*, 2019. Courtesy of EXIT International.
33	Lacy Barry, *Pillars of Fire & Cloud*, 2016. Photographer: Jennifer Endom. Medium: cardboard frame & paper. Size (approx.): 25 x 50 cm / 35 x 80 cm.
35	Nina van Bart, *Zooming In, Zooming Out	Tufting*, 2018.
37	Donald Baugh, *Roma collection of vessels, Pica*, 2019. Material: Walnut .
39	BC Biermann, *Venice x Venice*, 2019
41	Loreta Bilinskaite-Monie, *Journey Of the Thread*, 2018-19.
43	Felicia Björklund, *Light reflector (Highlights)*, 2012. Photographer: Mats Ringkvist.
45	Daniela Buonvino, *Rushes*, 2019.
47	Kevin Callaghan/Peter Carroll, *Precarious Earth I*, 2018
49	Barbara Calvo, *Narcissus' Revenge (round)*, Double-face hanging mirror, made up of three round elements, susceptible of being dismantled. Also available in other sizes and with different details. Size: cm 50 x 200 h x 6 + telescopic pole cm 50 x diameter 5.
51	Piero Castiglioni, *1954*, 2017.
53	Sophia Chraïbi Giorgi, *Papyrus*, 2019. Photographer: Clotaire Thomazo.
55	Yeongkyu Yoo, *Water(refill) Station*, 2019. Photographer: Romain Roucoules. Courtesy: CLOUDANDCO & COWAY.
57	Luce Couillet, *Halo N°2*, 2018. Photographer: Thomas Deron.
59	Creative Chef, Jasper Udink Ten Cate, *The Composition Table*, 2019. Photographer: Berend van Breda.
61	Dá Design Studio, *Danfo Std*, 2018.
63	Saverio D'Elia, *LUNENOTT*, 2019. Photographer: Stefano Cuccato.
65	DUODU, *COAT OF FEMAIL - omHAVNelser*, 2013. Photographer: Lasse Berre.
67	Mal Burkinshaw in collaboration with Sophie Hallette Lace, *Silhouettes en Dentelle*, 2013-2017. Photographer: Stuart Munro.
69	Yakusha Victoriya, *Set of vases KUMANTSY*, 2019.
71	Hicham Ghandour, *Wall light fixtures (detail)*, 2018. Photographer: Wael Khoury Photography.
73	Peter Ghyczy, *Jodie S02*, 1988.
75	Donna Glubo-Schwartz, *In Search of the Light Twisted Object*, 2018.
77	Julie Helles Eriksen, *Who are you?*, 2018.
78	Barrie Ho, Founder of BARRIE HO Architecture, *E-Sports Stadium @ ChongQing, China E-Sports Stadium @ ChongQing, China*, 2017.
80	Barrie Ho, Founder of BARRIE HO Architecture, *E-Sports Stadium @ ChongQing, China E-Sports Stadium @ ChongQing, China*, 2017.
82	Barrie Ho, Founder of BARRIE HO Architecture, *E-Sports*

	Stadium @ ChongQing, China E-Sports Stadium @ ChongQing, China, 2017.
85	Joca van der Horst, *Onda*, 2017.
87	Hsc Designs, *The Rocheux chair in Cement*, 2018. Photographer: Phx India.
89	Katalin Huszár, *Notjustuseless – pattern*, 2017. Photographer: Gergely Tarján.
91	Muraad Ibrahim, *Rose*, 2015.
93	Megumi Ito, *Lighting Object Kimono*, 2004. Photographer: Daniela Beranek.
95	Silvia Knüppel, *Neutralizer 01 / Household Sculptures*, 2019. Photographer: Tobias Bärmann.
97	Eva Levin, *Nature's raw form*, 2019. Photographer: Brian E. Barbarik. Courtesy: To Attila Kiraly for making it fly.
99	Benjamin Méry, *MINI Tryptic French Flag*, 2018. ©Lumneo.
101	MareikeLienauxUniqueFactory, *Pouf_Character_Trio*, 2018. Germany. Photographer: Benjamin Pritzkuleit.
103 105	Tahir Mahmood, *Rustam Dip Pen and Nib Holder Set*. MICAT, 2019.
107	Teresa Moorhouse, *Karhu*, 2019. Courtesy: Mum´s.
109	Brian Naeyaert, *This ain't New York, This the Bronx*, 2016. Photographer: Jeroom Vanderbeke. Courtesy of Bitch Studios.
111	Shrikant Nivasarkar, *PRESTIGE*.
113	Jinhwa Oh, *360° Alphabet from the front view, 0° rotation*.
115	Per Ploug, *Veizla Side Table and Cauda Chair*, 2017. Photographer: Andrew Miller Photography.
117	Yann Péron, *viséversa*, 2018-2019.
119	Estefania Johnson, *LUM*, 2019.
121	Revology Design Studio, *Where do you sit in nature? View over the Venetian Lagoon towards Chiesa di San Giogio Maggiore*, 2017. Photographer: Samuele Cherubini.
123	Lou van 't Riet, *triptych_1*, 2019.
125	Erdem Selek & Hale Selek, *Balance Mirror*, 2019
127	Gisela Simas, *Bia - chair and ottoman*, 2019. Photographer: Darwin Campos.
129	Mikaela Steby Stenfalk, *Process: Images collected from Instagram*, 2019.
131	Alexandra Kohl and J.M. Szymanski, *Lamp in Craved Iron and Cast Glass*, 2019.
132	Alexandra Kohl and J.M. Szymanski, *Bench in Craved Iron and Horse Hair Textile*, 2019.
133	Alexandra Kohl and J.M. Szymanski, *Bench in Craved Iron and Horse Hair Textile*, 2019.
135	Suzanne Tick, *Original process rendering for Woven Neon*, 2018.
137	Chantal Tramasure, *anhydre*, 2018.
139	Ritu Varuni, *Naga Roots Collection*, 2016-2019.
141	Wearable Media, *Sound Interactive Lightweight Coat, Color Gray*, 2018. Model: Estefania Rodriguez. Photographer: Christopher Zapata.
143	WORKTECHT & Namiko Kitaura, *BIOMBO - A Moment in a Fleeting World*, 2019. Photographer: Namiko Kitaura.
145	Jungmo Yang, *Gyeol chair*, 2018. ©Studio Jungmo Yang
147	Alexander Lorenz, *No 6*, 2019.
149	Monika Zabel, *Ode to the Ocean*, 2019. Photographer: Chris Lambertsen.
151	Hongtao ZHOU, Fang QI, Abaiyi-AKEBAI, Lianglu HOU, *Textscape-Tontsen Eye*, 2018. Image by: Xiaotong ZHANG, Hongliang WANG. Courtesy of: Lianglu HOU.
152	Hongtao ZHOU, Fang QI, Abaiyi-AKEBAI, Lianglu HOU, *Cloud of Measurement*, 2018. Photography by: Xiaotong ZHANG, Hongliang WANG. Courtesy of: Yumei QI, Kewei FENG, Jiabao ZHU.

SPONSORS

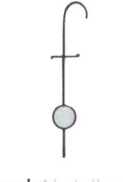